DEPRESSION AND SCHIZOPHRENIA

A Contribution on their Chemical Pathologies

H.M. van Praag, M.D., Ph.D.
Professor of Biological Psychiatry
Psychiatric University Clinic
Groningen, The Netherlands

S P Books Division of
SPECTRUM PUBLICATIONS, INC.
New York

to Seymour Kety

who inspired so many
to venture into the field of
biological psychiatric research

SPECTRUM PUBLICATIONS, INC.
175-20 Wexford Terrace, Jamaica, N.Y. 11432

Library of Congress Cataloging in Publication Data

Praag, Herman Meir van.
  Depression and schizophrenia.

  (Monographs in modern neurobiology)
  Bibliography: p.
  1. Depression, Mental--Physiological aspects.
2. Schizophrenia--Physiological aspect. 3. Neuropsy-
chopharmacology. 4. Amine metabolism. I. Title.
[DNLM: 1. Depression--Drug therapy. 2. Antidepres-
sive agents--Therapeutic use. 3. Schizophrenic--Drug
Therapy. 4. Tranquilizing agents--Therapeutic use.
QV108 P895d]
RC537.P68   616.8'95   76-49582
ISBN 0-89335-002-8

Preface

This book presents and discusses a number of hypotheses on the possible significance of disorders of the central monoamine metabolism in the pathogenesis of depressive and psychotic symptoms. Its character is largely clinical, in two ways. First, I have confined myself for the most part to clinical research done with regard to these hypotheses. Second, I have intended the book for all those who take an interest in the biological determinants of disturbed human behavior and biological methods of treating it.

In the past fifteen years, a substantial amount of work has been done concerning the relationship between disturbed behavior and monoamines; the pertinent literature, however, is unsurveyable and scattered through numerous journals. I felt the need to summarize it and "weigh" its merits, and I believed that the same need was felt by my colleagues who are interested in the subject. This was one of my reasons for writing this book.

There was also a second reason. This volume is an unfinished one, or rather, the theories discussed in it are incomplete. Numerous elements are still lacking, and others are uncertain. In fact, it may well be that much of what is asserted here will have to be retracted at some future time. But this would not detract from my intention to focus on an attempt to demonstrate that it is now possible to advance hypotheses on interrelations between disturbed cerebral function and disturbed behavior which, although largely derived from animal experiments, can be clinically tested. Until recently, this was an impossibility, a chimera. For this reason, this book is above all meant to be a touchstone of biological psychiatric thinking and research.

H.M. van Praag
Groningen, November 1975

Abbreviations

CA	catecholamines
CNS	central nervous system
COMT	catechol-O-methyltransferase
cyclic AMP	cyclic adenosine monophosphate
DA	dopamine
DBH	dopamine-β-hydroxylase
DOPA	dihydroxyphenylalanine
DOPS	dihydroxyphenylserine
5-HIAA	5-hydroxyindoleacetic acid
5-HT	5-hydroxytryptamine (serotonin)
5-HTP	5-hydroxytryptophan
HVA	homovanilic acid
IA	indoleamines
MA	monoamines
MAO	monoamine oxidase
methylene-THF reductase	5,10-methylene-tetrahydrofolate reductase
MHPG	3-methoxy-4-hydroxyphenyl glycol
a-MT	a-methyl-p-tyrosine
NA	noradrenaline
pCPA	p-chloro-phenylalanine
PEA	phenylethylamine
PIF	prolacting inhibiting factor
VMA	vanilylmandelic acid

Contents

PART II – SCHIZOPHRENIC PSYCHOSES

. . . However, one can be certain that all reductionist attempts to explain "mind" in terms of brain chemistry (only), or reinforcement schedules (only), or computer logic (only), or on any combinations of these, will fail. The facts of conscious experience are irreducible and must enter in their own right as basic irreducible elements, into any comprehensive account of mind.

J.R. Smythies
Psychological Medicine
August, 1973

PART I
Affective Disorders

Whereas medicine has always started from the practical aspect (for it is essentially an applied science), psychology and sociology have been largely theoretical in their origin. Their practical applications have come at the end rather than at the beginning of their work. Psychiatry must be regarded as one of the main branches of medicine. It is true that it has links with other sciences such as psychology and sociology but it is not unique in this respect, for this is equally true of preventive medicine. For this reason, psychiatry has much to learn from the methods of classification used in medicine, and the history of the development of these methods casts a ray of light on the present difficulties.

M. Hamilton
Classification in Psychiatry and Psychopathology,
National Institute of Mental Health, 1968

Introduction

In 1958, a new therapy for depressions was introduced: medication with antidepressants. Almost simultaneously, but independently, two types of compounds with an antidepressant effect were discovered: the tricyclic antidepressants (prototype = imipramine) and the MAO inhibitors (prototype = iproniazide). Previously, only amphetamine derivatives and opiates had been available for antidepressant medication. The effect of amphetamines on affective life was variable and brief, and often followed by a hangover; opiates actually acted mainly on anxiety and agitation. The antidepressants, with an effect on mood regulation which in suitable cases was primary and lasting, represented a new pharmacotherapeutic principle.

Although tricyclic antidepressants and MAO inhibitors are chemically unrelated, they proved to have two characteristics in common. In psychopathological terms, they exert a beneficial influence on depressions, particularly on a given syndromal type: the vital depression.* In biochemical terms, in the brain they behave like monoamine (MA) agonists, albeit via different mechanisms. This led to the hypothesis that the therapeutic effect of antidepressants is related to their ability to potentiate MA. This hypothesis was supported by the discovery that

---

*This syndrome is described in Anglo-American literature as endogenous depression. (Chapter1)

reserpine is a depressogenic substance, capable of provoking typical vital depressions, while it reduces the amount of MA available in the brain. Psychopathologically and biochemically, reserpine proved to be a counterpart of the antidepressants.

The suspected relation between antidepressant effect and MA potentiation raised the question whether depressive patients who show a favorable response to antidepressants are suffering from a functional deficiency in CA and/or 5-HT, or rather from a reduced susceptibility of postsynaptic CA and/or 5-HT receptors. In the past few years, this question has been approached from several different angles. Initially, the overall metabolism of MA was studied, there was not much else that could be done. Later investigations focused in particular on the CSF, and on post-mortem findings in the brain in suicide victims. Yet another strategy involved the pharmacological manipulation of the central MA metabolism, followed by studies to establish whether this manipulation relieves, aggravates or provokes depressive syndromes. An important product of this strategy was the development of drugs which can more or less selectively influence the metabolism of a given MA. Finally, efforts· were made to establish whether therapeutic methods with an established antidepressant effect, such as ECT, exert an influence on central MA.

Depression and mania, apparent antipodes, are often encountered in the same patients, and in psychodynamic views are closely related as to pathogenesis. It is therefore not surprising that biological depression research has also included manic patients, although their numbers have been small—probably because the manic patient does not readily yield to the exigencies of a strict research proto-- col.

The above discussed research led to two hypotheses: a CA hypothesis (Schildkraut, 1965; Bunney and Davis, 1965) and a 5-HT hypothesis (Van Praag, 1962; Ashcroft et al., 1965; Lapin and Oxenkrug, 1969; Coppen, 1967), which related the depressive syndrome to a central CA deficiency and 5-HT deficiency, respectively. Initially, these hypotheses stood side by side. The monistic view has now been virtually abolished, and few doubt that it is more meaningful to think in terms of disturbed balances. I shall henceforth confine myself to the MA hypothesis, by which I mean the hypothesis that disturbances in the central CA and/or 5-HT metabolism, leading to an absolute or relative deficiency in these substances, can play a role in the pathogenesis of depressive and manic syndromes.

The structure of the argumentation to be presented has been determined by the historical course of events as outlined above. To begin with, I shall discuss the influence of antidepressants and reserpine on the MA metabolism. Next I shall consider the arguments in favor of and against the MA hypothesis that result from studies in human individuals. I shall discuss a number of theories which have attempted to explain the suspected MA defect in disorders of affective life, and finally raise the question whether in these conditions indications have been found of defects in other central transmitter systems. The argumentation finally leads to an estimation of the "weight" of the MA hypothesis.

Prior to this argumentation I shall present a discussion of the diagnosis of depressions, which includes a discussion of the starting point of biological psychiatric research and one of the localization, metabolism and function of the central MA. The second theme is intended to deal with the misconception that biological research of behavior disorders has monopolist tendencies, and claims to trace "the" cause of these disorders while disregarding the importance of social and psychological determinants. The first and second themes have been added for those who are insufficiently familiar with the relevant psychiatric or neurobiological field. These sections are of an introductory character and are meant to elucidate the text which follows them.

CHAPTER I
Classification of Depressions

Multidimensional Diagnosis of Depressions

For many years, the pathological forms of depression (the distinction between physiological and pathological depressivity is left undiscussed here) were classified exclusively on the basis of etiological criteria. It was customary, and in some textbooks it still is, to refer to neurotic (psychogenic) depression, involutional depression, arteriosclerotic depression, endogenous depression, etc. (the list of "diagnoses" can be expanded ad infinitum because depressogenic factors are virtually innumerable). This seemed to suggest that a depression can be adequately characterized with its etiological adjective. Such a classification, however, is too superficial for adequate diagnosis, and adequate diagnosis is a prerequisite for the planning of both a therapy and a research protocol. In fact, this classification has become a source of confusion. Particularly in the fields of biological psychiatry and psychopharmacology, there have been investigators who have stressed the importance of adopting, in psychiatry also, some of the basic principles of medical nosology. They have insisted that (a) disease pictures, in this case depressions, should be classified according to their symptoms, course and causation, and (b) the concept "disease cause" should be divided into two components: etiology and pathogenesis (Van Praag and Leijnse, 1963b, 1965). Pathogenesis was defined as the constellation of cerebral dysfunctions which enable behavior disorders to occur; etiology was defined as all factors—hereditary, acquired somatic, psychological and social—which have contributed to the occurrence of the cerebral dysfunctions.

The following sections briefly discuss a number of aspects of the diagnosis of depressions (Table 1), and for the unwary (nonpsychiatric) reader three points should be made clear in advance.

TABLE 1

Classification of Depressions

Syndrome	Etiology		Course	Pathogenesis
Vital	Endogenous or Exogenous or Psycho(socio)genic or Idiopathic	Often combinations of these factors	Monophasic or Unipolar or Bipolar	Forms with and without disturbances in central MA metabolism
Personal	Psycho(socio)genic		Not well-known Recurrence is common	Unknown
Mixed forms	Not well-known Probably as in vital depression		Not well-known Recurrence is common	Unknown

(1) The multidimensional diagnosis of depressions advocated here is not general usage and is not employed in all publications.

(2) The terminology has not yet been standardized, primarily I am discussing the concepts which I myself use, but I shall attempt to "translate" them into the terms conventionally used in Anglo-American literature.

(3) The differentiation of depressive syndromes is still in its initial stage; the dichotomy discussed here is merely a preliminary attempt to chart the field.

Syndromal Points of View

From a strictly syndromal point of view, I distinguish two types of depression: vital and personal depression (Van Praag, 1962; Van Praag et al., 1965). Mixed forms are quite common; in other words, between the two prototypes there is a broad area which has not yet been adequately reconnoitered in terms of classification. Efforts to this effect are in full progress (e.g., Paykel, 1972; Garside et al., 1971, Becker, 1974). The following subsections describe what I regard as the cardinal symptoms of the *vital depressive syndrome*.

(1) *Despondency*. The mood is dejected in greater or lesser degree. The future seems obscure, and there are often marked suicidal tendencies. A typical feature is that the sense of comprehensibility is often absent: either no motive for the depression is experienced and the origin of the symptoms is an enigma to the patient, or an objective motive is evident in the history (e.g., the death of a loved one) but the patient does not experience this (any more) as an adequate explanation. The comprehensible correlation between motive and depression has been lost. This is the case in the so-called vitalized personal depression (see below).

(2) *Motor Disorders*. In principle, these carry a minus sign (retarded vital depression). In severe cases they are objectively observable (in facial expression and gestures), and the retardation is apparent also in speech and thinking (slow thought processes, impoverished dialogue, monosyllabic answers). In the (quite common) milder cases, the retardation is apparent only in subjective experiencing. The patient finds that his work is no longer going smoothly and that he cannot cope with it unless he makes a determined effort. Nothing comes easily any more. Ideas are reluctant to come and can be retained only with difficulty. The patient himself speaks of disturbed concentration. In some cases there may be agitation at the same time (agitated vital depression): the patient feels tense and anxious, and may show manifest motor unrest. I have said "at the same time": in these cases retardation and agitation are not polar factors, but the agitation is superimposed on the state of retardation.

(3) *Hypoaesthesia*. The ability to respond emotionally to exogenous stimuli is diminished. Things from which the patient used to derive pleasure—his work, his family, his hobbies—no longer mean much to him. He can no longer enjoy things. He is listless in the true sense of the word. In a later stage, his ability to feel grief also diminishes. Hypoaesthesia is not to be equated to depersonalization. In the former, emotional experiencing is quantitatively reduced; in the latter, it is qualitatively changed.

(4) *Somatic Disorders*. An obligatory triad, in my opinion, is that of disorders of sleep (early-morning awakening), fatigue disproportionate to performance, and reduced appetite. The actual amount of food taken may still be normal, but it no longer tastes good. Sexual disorders (loss of libido, impotence, amenorrhoea) are common features but not, in my view, essential to the diagnosis. In some cases, somatic complaints are hypochondriacally exaggerated and given the focus of attention. This entails the risk of failure to recognize their origin (the depression); the resulting condition is a so-called masked depression.

(5) *Rhythmicity*. The symptoms are most severe upon awakening and in the morning hours, and show some improvement in the course of the day. In more severe cases, this fluctuation diminishes and finally disappears completely, the patient being despondent throughout the day. Seasonal influences (appearance in spring and/or autumn) are classical but by no means evident in all cases.

In severe vital depressions, delusions can develop: in particular, delusions of guilt, sin, poverty and hypochondriacal complaints. For these cases I use the designation *melancholia.*

In the following subsections the cardinal symptoms of the *personal depressive syndrome* will be outlined as parallels of the features of the vital depressive syndrome.

(1) In these cases, too, the patient is despondent, but the mood anomaly is comprehensible to him in that there is always a motive which is experienced as an adequate explanation of the gloomy mood.

(2) Disorders of motor activity and intrinsic motivation are much less pronounced. Significant retardation is rare, but agitation and anxiety are common.

(3) There is no hypoaesthesia, but depersonalization may occur.

(4) Somatic disorders are common, but mainly involve expressions of autonomic dysregulation such as palpitations, atypical anginous pains, disagreeable sensations at various sites in the body, etc. Sleep is often disturbed, particularly onset of sleep and continued sleep.

(5) There is no circadian rhythm of the type observed in vital depression.

The syndrome of personal depression is less clearly defined than that of vital depression, because there are transitions to the vast field of the psychosomatic diseases, functional somatic disorders, anxiety neuroses, etc. As long as this is not more accurately charted, the description of a personal depression cannot be more exact.

Roughly, the vital depressive syndrome corresponds with the syndrome described in Anglo-American literature under the heading *endogenous depression,* while the personal depressive syndrome corresponds with what is described under such headings as *neurotic, reactive* or *psychogenic depression.* What I call melancholia would seem to me to be identical with the so-called "psychotic depression."

Some authors have maintained that the difference between vital (endogenous) and personal (neurotic) depression is exclusively quantitative (Becker, 1974): personal depressions are alleged to be relatively mild, whereas vital depressions are always severe. I do not share this view. In my opinion, the difference between the two types of depression is a qualitative one, and either type can occur in a mild or severe form.

Etiological Points of View

In the context of this approach, an estimate is made of the extent to which endogenous (hereditary), exogenous (acquired somatic), psychogenic (intrapsychic) or sociogenic (relational and environmental) factors have contributed to the development of a given syndrome.

In the individual case, the sole indication of the activity of an endogenous factor is a tainted family history, and the taint should not be exclusively apparent in the direct line because in that case pseudo-heredity (transmission via a learning process) cannot be eliminated. Exogenous factors are investigated on the basis of the history and physical examination.

A psychogenic "load" becomes plausible when there are positive indications of an unresolved mental conflict and a personality structure with weak spots which make it understandable why the conflict remained unresolved (Chodoff, 1972).

Sociogenic stress factors have their origin in the patient's life environment (Brown et al., 1973). However, their pathogenic action is always a resultant of the interaction between psycho-trauma and personality structure. Most people can independently cope with the majority of their frustrations. A minority show decompensation. There should therefore be demonstrable features in the personality structure which explain its "susceptibility."

The vital depression is an etiologically nonspecific syndrome which can be caused by endogenous as well as exogenous or psychosocial factors. The vital depression of chiefly endogenous determination is identical to the classical endogenous depression in conventional psychiatric literature. I prefer the former designation because (1) it is quite possible that an endogenous factor also plays a role in nonvital depressions, and (b) vital depression can also be provoked by nonendogenous factors. Virus infections such as influenza, infectious mononucleosis and infectious hepatitis are an example of an exogenous causation of vital depressions. The nonpsychiatrist notes that these diseases have a prolonged period of convalescence. Psychiatrically speaking, these periods are often characterized by a mild vital depressive syndrome. Vital depressions can also be psychogenic and sociogenic (Van Praag, 1962). The depression can have a vital character from the onset or initially be of a personal character and gradually change into a vital depression: the depression becomes "vitalized."

In many actual cases, combinations of factors are involved. For example, an individual living under stress, with a depressively tainted family history, develops a vital depression after experiencing a physical illness. Finally, a vital depression can develop without any discernible etiology; the term *idiopathic* applies to such cases.

In the origin of the personal depressive syndrome, psychogenic and sociogenic factors invariably play an important role. In fact, this syndrome corresponds with that of the psychogenic and neurotic depression in the "official" psychiatric literature. I prefer the designation *psychogenic* (or *neurotic*) personal depression, because psychogenic influences can also lead to a vital depression.

A vital depression is a self-limiting disease which can last from a few days to many years but tends toward spontaneous recovery. It is as if, once developed, the depression takes an independent course following its own rules, regardless of the causative factors. Reserpine depression provides an excellent example. Discontinuation of reserpine usually does not ensure recovery. Antidepressant treatment is required to achieve this. For this reason one may pose the questions: Is it correct to distinguish exogenous and psychogenic vital depressions? Is not rather every vital depression essentially endogenous, with the exogenous and psycho(socio)genic factors merely subordinate, precipitating elements? This question is unanswerable but, in view of recent biochemical research (see below), I regard this possibility as by no means excluded.

Classification According to Course

In terms of their clinical course, two types of depression are distinguished: unipolar and bipolar. A unipolar depression is defined as a recurrent vital depression without (hypo)manic phases. A bipolar depression is a recurrent vital depression in the course of which (hypo)manic phases also occur. Unipolar and bipolar depressions differ in other features as well. The mode of hereditary transmission is not the same (it is to be understood that genetic factors are by no means always demonstrable), but there are differences also in age of onset of the first phase, the duration of symptom-free intervals, and the premorbid personality structure (Perris, 1966, Angst, 1966).

The terms *unipolar* and *bipolar* hold no etiological implications. In the classical cases, the phases develop for no apparent reason and the family history is depressively tainted; they show the classical endogenous etiology. However, cases in which the phases are provoked (or caused?) by exogenous and/or psychosocial factors are far from exceptional, particularly in the earlier stage. Not infrequently, later phases tend to develop more and more spontaneously.

The term *manic-depressive psychosis* can be regarded as a collective term for unipolar and bipolar depressions. I am disinclined to use this designation because (a) a majority of patients never develop (hypo)manic phases, and (b) a majority of patients never become psychotic either in their depressive or in their (hypo)-manic phases.

If left untreated, the psychogenic (neurotic) personal depressive syndrome also tends toward recurrence. This is to be expected. As long as the patient's personality structure and life situation fail to change, there is a continued risk of minor or major conflicts which can result in a depression. Unlike the natural course of the vital depression, that of the personal depression has so far been given hardly any systematic investigation.

I personally confine the term unipolar depression to recurrent vital depressions. In the literature, it is by no means always clear whether or not this definition is accepted and used. In such cases, the term unipolar (one-dimensional as it actually is) is confusing.

Pathogenetic Points of View

The introduction of the pathogenesis concept in psychiatry (Van Praag and Leijnse, 1963b, 1965) implies the assumptions that (a) the disease concept is of a material nature in psychiatry also, and (b) every state of behavior, disturbed or undisturbed, is dependent on a given functional state of the brain—a state which in principle can be analyzed and influenced by biochemical (drugs) or physiological (depth electrodes) means. To put it briefly: there is no disturbed behavior without corresponding cerebral substrate. This statement is not a revival of nineteenth-century materialism. It does not hold that behavior disorders *are* diseases of the brain. What it does hold is that pathogenic influences of any kind—be they psychological, environmental or somatic—affect mental life not directly, via some sort of vacuum, but indirectly, via changes in cerebral organization. In this view, the brain takes the role of an intermediary agent.

Anyone who accepts this view thereby declares himself an opponent to the way of thinking in alternatives, which is still common usage in psychiatry. An example: "Is behavior disorder A a biochemical or a psychosocial disease?" In my opinion, this question poses a spurious problem. If a dichotomy is at all required, then the separation should be made as follows (Fig. 1):

(1) Behavior disorders in which psychological and social factors contribute in an important degree to the development of the cerebral dysfunctions which generate these disorders.

(2) Behavior disorders based on cerebral dysfunctions which to an important extent must be ascribed to factors other than psychosocial factors: e.g., acquired somatic diseases which involve damage to the brain, and hereditary factors due to which, say, a given enzyme primordium is marginal.

Another example: "Behavior disorder A can only be treated by psychotherapy, and behavior disorder B responds only to pharmacotherapy." This statement also contains an inconsistency of logic. The purposes of pharmacotherapy and psychotherapy (here meant to include sociotherapy) are disparate. Drugs are resorted to in an effort to control cerebral dysfunctions. They are aimed at the substrate which generates the disease symptoms. Psychotherapy is aimed at the pathogenic input. Its purpose is to attenuate tensions generated within the individual or by the interaction between individual and environment. In other words, the objectives of pharmacotherapy and those of psychotherapy are quite different. The former aims at the pathogenesis, the latter at the etiology, of the syndrome presented. As such, they are complements and must be applied jointly, at least insofar as adequate methods are already available for a given diagnostic category.

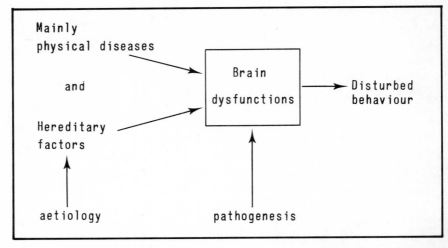

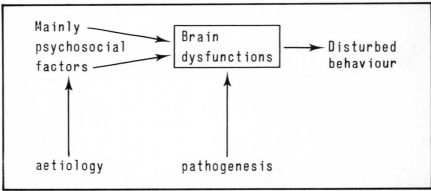

1 Relationship between biological and environmental factors in the occurrence of behavioral disorders.

A pathogenetic classification of depressions is based on data of cerebral function. Research on the basis of this criterion is still in its initial stage. Preliminary results, however, are encouraging in that they tend to show that this approach is not based on fiction. I shall not discuss these findings here, for they are the terminal point and the quintessence of the arguments presented in subsequent chapters.

Other Diagnostic Classifications

Chiefly on the basis of life history and the premorbid personality structure, Robins et al. (1972) made a distinction between primary and secondary affective disorders. The primary depression occurs in individuals with a more or less normal personality structure and with a psychiatric history which is "clean," apart from possible depressions or manic phases. The depression is called secondary in patients with preexistent psychiatric disorders other than depression and mania, and with a disturbed personality structure.

The primary depression concept in fact comprises the classical vital depression with a unipolar or bipolar course in non-neurotic individuals; the secondary depression concept covers the neurotic depression, but gives insufficient information on the symptomatology. I am not too happy with these two concepts. They are not multidimensional, they ignore the possibility that vital depressions (with a unipolar or bipolar course) can also occur in neurotic individuals, and there are no indications that predominantly neurotic and predominantly non-neurotic vital depressions are fundamentally different in therapeutic or prognostic terms.

On the basis of an exhaustive family study, Winokur et al. (1971) distinguished two subtypes within the unipolar group: depression spectrum disease and pure depressive disease. A prototype of the former group is a woman whose first depressive phase developed before age forty, and in whose family depression is more common among women than among men, with the "depression deficit" in the male line "repleted" by alcoholism and sociopathy. A prototype of the latter group is a man whose first depressive phase started after age forty and in whose family depressions are as common in men as in women, while alcoholism and sociopathy are not overrepresented.

No biochemical research has yet been done on the basis of the latter classification.

Conclusions

In this chapter, a plea was presented in favor of a multidimensional classification of depressions (and other psychiatric categories) according to symptomatology, etiology, course and pathogenesis. This approach is a necessity because (a) the nature of the syndrome gives insufficient information on its etiology, (b) a given etiology does not necessarily lead to a given syndrome, and (c) the course of a depression cannot be reliably predicted either on the basis of its etiology or on that of the syndromal features. A multidimensional approach is of importance for clinical practice; it facilitates therapeutic planning. Schematically and by way of example: the decision as to "predominantly vital" or "predominantly personal" depression determines whether antidepressants will or will not be given; the decision as to the "weight" of the psychosocial factors determines whether focused psychotherapy will be resorted to, and the decision as to "unipolar" or "bipolar" depression determines whether lithium prophylaxis will be used.

It seems to me that a multidimensional approach is indispensable for research purposes.

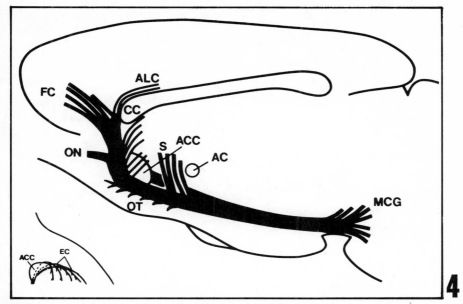

4

Schematic representation of the presumed dopaminergic mesolimbic and mesocortical systems in a sagittal projection from Lindvall and Bjorklund, 1974.
  AC = anterior commissure; ACC = nucleus accumbens; ALC = anterior limbic cortex; CC = corpus callosum; FC = frontal cortex; MCG = mesencephalic CA cell groups; ON = olfactory nuclei; OT = olfactory tubercle; S = septum.

amount of DA in the brain. The mesolimbic system ranks a fair second with about 20-25%.

The cell bodies containing 5-HT are to be found chiefly in the raphe nuclei localized in the mesencephalon and the cranial part of the medulla oblongata (Dahlström et al., 1973, Lorens and Guldberg, 1974). The latter nuclei (nuclei raphe pallidus and obscurus) extend fibers to the medulla oblongata and spinal cord. The mesencephalic nuclei (nuclei raphe dorsalis and medianus) extend fibers rostradly to the hypothalamus, corpus striatum, neocortex (particularly frontal and parietal cortex) and many other areas. 5-HT terminals are found in varying density at numerous sites in the brain.

Cell bodies containing DA and 5-HT thus form fairly well-defined "nuclei" which are relatively readily accessible to techniques of traumatization and stimulation. This has greatly facilitated the study of the function of these systems.

MA-ergic Transmission in the CNS

Impulse transmission in a MA-ergic synapse is probably effected as follows: (Fig. 5, Fig. 6)

(1) An impulse reaches the receptive part of a nerve cell and generates an action potential by depolarization of the axon membrane. The action potential is conducted to the nerve ending and, in response to the depolarization process, transmitter substance is released into the synaptic cleft.

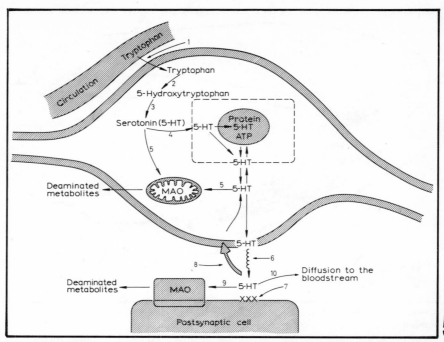

Schematic representation of serotonergic synapse (modified after Cooper et al., *The Biochemical Basis of Neuropharmacology,* Oxford University Press, London, 1974).

1: Active uptake or transport of tryptophan in the CNS; 2 and 3: biosynthesis of 5-HT; 4: uptake of 5-HT in synaptic vesicles and binding with ATP and protein; 5: degradation of unbound ("free") 5-HT by the enzyme MAO, localized largely in the outer membrane of mitochondria; 6: release of 5-HT in the synaptic cleft; 7: interaction of the transmitter with the postsynaptic receptors; 8: re-uptake of 5-HT in the synaptic vesicles; 9: extraneuronal degradation of 5-HT, presumably by MAO (extraneuronal localization unknown); 10: diffusion of 5-HT to the circulation.

In principle, the transmitter can originate from the cytoplasmatic or from the vesicular fraction. In the former case, we must assume either that the presynaptic membrane's permeability for the transmitter increases in response to the impulse, or that the binding of the transmitter to ATP becomes less stable, as a result of which the transmitter concentration in the cytoplasm increases and leakage to the synaptic cleft occurs. This, however, is improbable in view of the results of histochemical studies which cannot be discussed in detail here (Farnebo and Hamberger, 1971). This implies that the transmitter must originate from the synaptic vesicles. On this point there are three theories: (a) the vesicle is released in its entirety to the synaptic cleft, where it disintegrates; (b) the vesicle comes into contact with the presynaptic membrane and an aperture forms at the site of contact between the two membranes, whereupon the contents of the vesicle are released into the synaptic cleft—a process known as exocytosis; (c) a synaptic vesicle releases its contents in the immediate vicinity of the presynaptic membrane, and the transmitter diffuses through this membrane into

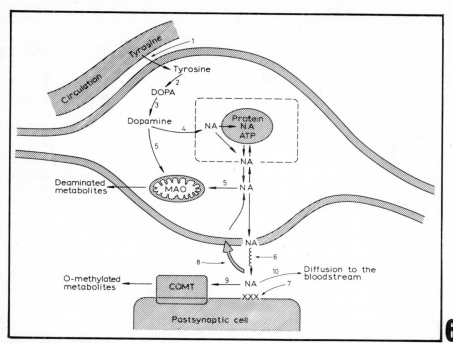

Schematic representation of a noradrenergic synapse (modified after Cooper et al., *The Biochemical Basis of Neuropharmacology*, Oxford University Press, London, 1974)

1: Active uptake or transport of tyrosine in the CNS; 2: and 3: biosynthesis of DA; 4: uptake of DA in synaptic vesicles, transformation in NA and binding with ATP and protein; 5: degradation of unbound ("free") DA and NA by the enzyme MAO localization largely in the outer membrane of mitochondria; 6: release of NA in the synaptic cleft; 7: interaction of the transmitter with the postsynaptic receptor; 8: re-uptake of NA in the synaptic vesicles; 9: extraneuronal degradation of NA, presumably largely by COMT; 10: diffusion of NA in the circulation.

In a dopaminergic synapse DA-β-hydroxylase is not available. Therefore, DA is not transformed into NA, but stored as such. For the rest, structure and functioning of both types of synapses are considered to be similar.

the synaptic cleft. The last-mentioned mechanism is the least plausible. In the periphery, after excitation of a NA-ergic nerve, not only is NA found extra-neuronally, but also other vesicle constituents such as ATP, chromogranine and certain enzymes (De Potter et al., 1969). In view of the dimensions of these molecules, their ability to pass through the presynaptic membrane is questionable. On the basis of morphological and histochemical findings, exocytosis is now being regarded as the most plausible release process, although there are some indications that there can be extraneuronal (synaptic?) vesicles also (Grillo, 1970).

The mechanism by which depolarization of the axon membrane causes migration of synaptic vesicles to the presynaptic membrane, and the exact mechanism of exocytosis, are not well known. The influx of calcium ions from the extracellular space into the nerve ending probably plays a role in this respect (Smith and Winkler, 1972).

(2) Transmitter substance released into the synaptic cleft diffuses to the postsynaptic membrane, where it binds with the postsynaptic receptors. Via a mechanism as yet unknown, this leads to a change in the ion permeability of the postsynaptic membrane, Cyclic adenosine monophosphate (cyclic AMP) possibly plays a role in this respect, for CA activate adenylcyclase in the postsynaptic element, and this is the enzyme which converts ATP into cyclic AMP. Local accumulation of cyclic AMP leads to phosphorylation of membrane proteins, and this might alter the ion permeability of the membrane.

Be that as it may, when the transmitter increases the sodium ion permeability, the membrane depolarizes and the postsynaptic element becomes subject to a state of excitation which is conducted on to the axon. The reverse happens when the transmitter increases the chloride ion and possibly the potassium ion permeability. The postsynaptic system is then hyperpolarized: it is less readily excitable, or inhibited. Biochemists and biophysicists are now focusing considerable effort on isolation and identification of postsynaptic receptors.

(3) Once the signal has been transmitted to the next neuron, the transmitter substance disappears from the synaptic cleft, partly by extraneuronal degradation and diffusion to the blood stream, but for the most part by re-uptake into the nerve ending and hence into the synaptic vesicles. This (re-)uptake also takes place against a concentration gradient. For example, brain slices concentrate NA from an incubation medium which has a 5-8 times higher NA concentration. This means that an active, energy-consuming process is involved: a true pump mechanism.

As already pointed out, the transmitter function of MA in the CNS—although quite plausible—has not been demonstrated with certainty. A transmitter, and the enzymes involved in its synthesis and elimination, ought to be localized in the presynaptic element of a synapse. The MA meet this criterion. A transmitter ought to be released upon excitation of a nerve, and, guided to the postsynaptic receptor, to produce the same effect as excitation of the presynaptic nerve fiber. Owing to the enormous complexity of the central neuronal network, it has not been established with certainty whether MA meet these criteria. Release processes have been studied by such means as the use of a so-called "push-pull" cannula and by perfusion of the lateral ventricles. For the study of the postsynaptic effect of MA, micro-iontophoresis has proved to be a valuable aid. It is in view of results thus obtained that the transmitter function of MA is plausible. But these results are not conclusive, for the techniques used merely give an impression of events which take place in groups of neurons. They are far too crude to give information on events within a given individual synapse.

Metabolism of CA

It is difficult for CA to enter the brain from the bloodstream. They must therefore be *produced* (Fig. 7). The mother substance is the amino acid tyrosine. which, via an active transport mechanism, is taken up into the brain and then into CA-ergic neurons. Several aromatic amino acids compete for the same transport system, and consequently the uptake of tyrosine (and tryptophan, mother substance of 5-HT) into the brain diminishes in the case of, say, phenylketonuria, with high plasma phenylalanine levels. Tyrosine is involved in numerous metabolic reactions, and only a very small amount is utilized for CA synthesis (Fig. 7). For this purpose, tyrosine is hydroxylated at the 3-site by the enzyme tyrosine hydroxylase, in which process DOPA is produced. This reaction is speed-limiting in the synthesis of CA. With the aid of DOPA decarboxylase, CO_2 is withdrawn from DOPA, and DA results. Tyrosine hydroxylase has a high substrate specificity, but DOPA decarboxylase has not: it decarboxylates all natural aromatic l-amino-acids such as histidine, tyrosine, tryptophan and phenylalanine, as well as 6-HTP and DOPA. A more appropriate designation for this enzyme would therefore be l-aromatic amino acid decarboxylase. Next, DA is oxidized in the side chain by DA-β—hydroxylase, and NA is formed in this process. DA-β-hydroxylase is localized in the synaptic vesicles, whereas the other enzymes are contained in the cytoplasm of the nerve endings.

The enzyme phenylethanolamine-N-methyltransferase, which converts NA to adrenaline via N-methylation and which is found in abundance in the adrenal medulla, occurs also in the mammalian brain, particularly in areas involved in the olfactory functions (olfactory bulbus and olfactory tubercle); however, its activity is low and, accordingly, the adrenaline concentration. Moreover, the function of central adrenaline is unknown.

CA are *degraded* by oxidation and methylation, under the influence of monoamine oxidase (MAO) and catechol-O-methyltransferase (COMT). (Fig. 8) Contrary to what is often contended, the former enzyme is not found exclusively intraneuronally, bound to the mitochondrial membrane, but also outside neurons: after denervation, the MAO activity in the sympathetically innervated end organ shows only a moderate decrease. It is plausible, on the other hand, that mainly intraneuronal MAO is involved in MA degradation. COMT functions chiefly extraneuronally, although its exact localization remains obscure. MAO converts CA to aldehydes, which are immediately further degraded either to the corresponding acid (by the enzyme aldehyde dehydrogenase) or to the corresponding alcohol or glycol (by the enzyme aldehyde reductase). It has recently been demonstrated with the aid of electrophoresis that MAO is not a simple enzyme but occurs in at least two different types. These are called isoenzymes, but, strictly speaking, this is an inappropriate name because they differ in substrate specificity (which means that a given type of MAO acts on a given MA and is less active in relation to other MA), and susceptibility to MAO inhibitors. In fact, they are different (if related) enzymes.

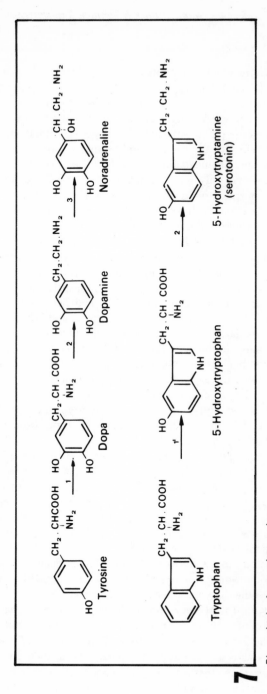

Biosynthesis of transmitter amines.

1=tyrosine hydroxylase; 2=DA (5-HT) decarboxylase; 3=DA-β-hydroxylase;
1¹=tryptophan hydroxylase.

7

26

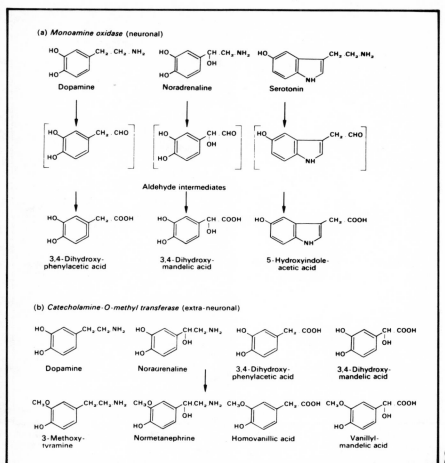

(a) *Monoamine oxidase* (neuronal)

Dopamine   Noradrenaline   Serotonin

Aldehyde intermediates

3,4-Dihydroxy-
phenylacetic acid

3,4-Dihydroxy-
mandelic acid

5-Hydroxyindole-
acetic acid

(b) *Catecholamine-O-methyl transferase* (extra-neuronal)

Dopamine   Noradrenaline   3,4-Dihydroxy-
phenylacetic acid   3,4-Dihydroxy-
mandelic acid

3-Methoxy-
tyramine   Normetanephrine   Homovanillic acid   Vanillylmandelic acid

8

Degradation of 5-HT, DA (peripheral and central) and NA (peripheral) (from H.S. Bachelard, *Brain Chemistry*, Chapman and Hall, London, 1974).

COMT catalyzes the transmission of a methyl group from the methyl donor S-adenosylmethionine to the 3-hydroxyl group of the CA. In the CNS, the principal degradation product of NA is a glycol: 3-methoxy-4-hydroxyphenyl-glycol (MHPG) (Fig. 9); this is in contrast to the periphery, where NA is largely converted to an acid (vanillylmandelic acid).

Human central DA is chiefly converted to an acid: homovanillic acid (HVA). The corresponding alcohol (3-methoxy-4-hydroxyphenylethanol; MOPET) is quantitatively of subordinate importance.

Major catabolic pathways of NA in the CNS. MAO=monoamine oxidase; AlcDH=alcohol dehydrogenase; COMT=catechol-O-methyltransferase.

Metabolism of 5-HT

For the production of 5-HT, too, the brain must rely on itself: it is difficult for this amine to enter the brain from the bloodstream. The mother substance of 5-HT is the essential amino acid tryptophan (Fig. 7). This is first of all hydroxylated to 5-HTP. The activity of the enzyme tryptophan-5-hydroxylase, which is involved in this, is low in the brain, and this has greatly impeded its detection and identification. It is a cytoplasmic enzyme of high specificity—different from, say, tyrosine hydroxylase and phenylalanine hydroxylase (which forms tyrosine)—which probably occurs exclusively in serotonergic neurons.

Next, 5-HTP is decarboxylated to 5-HT with the aid of 5-HTP decarboxylase—an enzyme of low specificity which is identical to DOPA decarboxylase.

Given the presence of a sufficient amount of tryptophan, conversion of trypto-phan to 5-HTP is the speed-limiting factor in 5-HT synthesis.

The principal route of 5-HT degradation is: deamination to 5-hydroxyin-doleacetaldehyde in response to MAO, and further oxidation to 5-hydroxyin-doleacetic acid (5-HIAA) with the aid of aldehyde dehydrogenase (Fig. 8). In principle, the aldehyde can also be reduced to 5-hydroxytryptophol. Whether this conversion is of any importance in the CNS is unknown. The brain contains enzymes able to convert 5-HT directly (without deamination) to the 5-sulfate ester. Under normal conditions, however, this reaction is probably of minor importance.

False Transmitters

DOPA (= 5-HTP) decarboxylase and DA-β-hydroxylase have a low substrate specificity. Consequently, numerous compounds capable of being taken up into the MA-ergic cell are converted by these enzymes and then taken up into the synaptic vesicles. In this way, they can come to function as a so-called "false transmitter": in response to an impulse they are released into a synaptic cleft, but they are incapable of exciting the postsynaptic receptor. An example (in the course of this argumentation, we will encounter several): when a large amount of l-DOPA is administered, part of it is taken up into serotonergic neu-rons and converted by DOPA (= 5-HTP) decarboxylase to DA, which in these neurons probably starts to function as a false transmitter. Another example: α-methyldopa (Aldomet, a hypotensive) is converted by DOPA decarboxylase and DA-β-hydroxylase to α-methyl-DA and α-methyl-NA: potential false trans-mitters in DA-ergic and NA-ergic neurons, respectively.

Regulation of the Release and Synthesis of MA

Many publications have been devoted to this subject in the past few years. Since it is impossible to give a concise outline of this literature, I shall merely mention a few possibilities. Whether they are all really active as regulatory mechanisms, and, if so, under which conditions, is uncertain (Hamon and Glowinsky, 1974; Carlsson et al., 1972).

The amount of NA released into the synaptic cleft is probably regulated via α-receptors in the presynaptic membrane. These are believed to be excited when the NA concentration in the synaptic cleft exceeds a given critical level, with, as a result, cessation of the release process. Whether a similar mechanism also exists in the case of the other MA is unknown.

An obvious possibility for the regulation of transmitter synthesis is so-called "end product regulation." CA inhibit tyrosine hydroxylase activity *in vitro*. It is therefore assumed that an increase in the concentration of CA in the neuron inhibits their synthesis, and vice versa. In view of the probably inconsiderable changes in the concentration in the nerve endings, it is doubtful whether this mechanism is active under normal conditions.

Another synthesis-regulating factor is the impulse flow in the presynaptic system. This has long been known with regard to NA. In the peripheral sympathetic nerves, the NA concentration is constant within narrow limits, and independent of neuronal activity over a wide range. Consequently, there must be a homoeostatic mechanism which keeps the transmitter concentration constant, even though a fraction of this is lost at each impulse transmission. It is likely that the activity of aromatic amino acid hydroxylases (tryptophan hydroxylase and tyrosine hydroxylase) is indeed influenced by nerve impulses. Whether this is a direct influence or an indirect one, e.g., via the availability of a cofactor, is unknown.

Another possibility is that postsynaptic receptors exert an influence on transmitter synthesis. The existence of such a mechanism became a probability when it was found that excitation of postsynaptic CA and 5-HT receptors reduces the synthesis of the corresponding transmitter, and vice versa. The manner of the transmission of such a signal, if any, has remained obscure. No nerve pathway which can serve this purpose has been identified.

Finally, there can be little doubt that central MA synthesis is subject to hormonal influences as well. For example, the conversion of ^{14}C-labeled tyrosine to ^{14}C-labeled NA is increased following thyroidectomy. This might well be an important future crossroads for a meeting of transmitter chemistry and (neuro)-endocrinology.

Measuring MA Turnover

The term "turnover" refers to the overall rate at which the entire amine store available in a given tissue is replaced. It is a chemical concept, and supplies no information on the question of whether all the amine converted has been utilized for transmission. The turnover rate of an amine must therefore not be regarded simply as an indicator of the functional condition of the system. On the other hand, there are indications that the neuronal activity in MA-ergic neurons is a factor of importance in regulating synthesis. For this reason, it is often hypothesized that the reverse also applies: chemical turnover is indeed an index of the functional activity in the system. This, however, remains to be proven.

There are several methods of measuring central MA turnover; none of them is ideal. First of all, there are methods which make use of a drug. We can inhibit tyrosine hydroxylase with the aid of α-MT, and tryptophan hydroxylase with pCPA; the central concentration of CA and 5-HT, respectively, then begins to diminish, and the rate of disappearance is a measure of the turnover of these substances. It is also possible to inhibit MA degradation (with the aid of a MAO inhibitor), and then to measure the accumulation of the amines as an indicator of their rate of synthesis. Finally, there is the probenecid technique. Probenecid is a substance that inhibits the transport of the acid 5-HT and DA metabolites 5-HIAA and HVA from the CNS to the bloodstream. As a result, they accumulate in the brain and CSF, and the rate of accumulation equals the rate of their synthesis (at least for a few hours). The rate of synthesis of 5-HIAA and HVA and the rate of degradation of 5-HT and DA are of identical magnitudes. The rate of accumulation of 5-HIAA and HVA after probenecid administration is therefore a measure of the rate of degradation of the mother amine.

For the purpose of these studies, the above-mentioned drugs must be given in large doses, in which they produce unintended effects as well. This is a disadvantage. The former two methods, moreover, lead to marked changes in MA concentration, and it is conceivable that this, as such, could influence the normal control mechanisms.

Radioisotope techniques have also been used in measuring MA turnover. To begin with, labeled amine is introduced intracisternally or into the ventricle (intravenous administration is impossible because MA do not readily pass the blood-brain barrier), whereupon the rate of disappearance of the specific activity of the amine in question is measured. This method starts from three assumptions: (a) that the labeled amine is selectively taken up by the corresponding MA-ergic neurons, (b) that the tracer dose of the isotope is so small that the endogenous amine pool is not or hardly enlarged; (c) that the amine lies stored in a homogeneous pool. The fact that these requirements are not entirely met reduces the validity of this method.

An alternative radioisotope technique calls for intravenous injection of labeled precursors—e.g., tyrosine for measuring CA turnover and tryptophan for measuring 5-HT turnover—whereupon the decrease in the specific activity of the precursor and the increase in that of the amine are measured. This technique, too, has its flaws. For example, it assumes that the MA-ergic neuron has no preference for freshly formed amine over that of the old store for transmission; in fact, however, there are indications that it has such a preference (Schildkraut et al., 1971; Glowinski et al., 1972).

Conclusions

MA probably play a role as transmitters in a number of fairly well-defined groups of neurons in the CNS, localized in areas traditionally assumed to figure in the generation and regulation of motor activity (basal ganglia), emotionality and level of motivation (limbic system). This explains the great interest in these compounds displayed by investigators trying to fathom the relations between brain and behavior.

CHAPTER III
Tricyclic Antidepressants and Central Monoamine Metabolism

Tricyclic antidepressants are assumed to increase the 5-HT and NA concentrations at the corresponding cerebral receptors. The activities in these systems are thought to increase as a result. The increase in MA concentration is conceived of as being caused by inhibition of the membrane pump which ensures that 5-HT and NA are returned from the synaptic cleft through the membrane of the neuron to the synaptic vesicles.

I shall now discuss the arguments in favor of the hypothesis that tricyclic antidepressants are inhibitors of the central 5-HT and NA "pump." They tend to show the plausibility of inhibition of uptake, not inhibition of re-uptake. The last-mentioned process is not yet measurable. It seems useful to establish this in advance.

Tricyclic Antidepressants and NA Uptake

Potentiation of peripheral and central NA effects. Tricyclic antidepressants potentiate several peripheral effects of exogenous NA, e.g., increase in blood pressure and contraction of the feline nictitating membrane (Sigg, 1959; Cairncross, 1965; Eble, 1964; Theobald et al., 1964). Potentiation of the effect on blood pressure has been observed in human subjects also (Prange et al., 1964; Fischbach et al., 1966). The effect of endogenous NA, i.e., NA released from

33

sympathetic nerve endings after excitation, is likewise potentiated by tricyclic antidepressants, e.g., the response of the feline nictitating membrane (Sigg, 1959, Cairncross, 1965; Haefely et al., 1964).

The question of whether tricyclic antidepressants potentiate intracerebral NA as well cannot be immediately answered because we have no sufficient exact information on central NA effects. On indirect grounds, however, it seems plausible: tricyclic antidepressants, after all, potentiate effects of substances whose action is related to increased NA activity. For example, they potentiate various effects of amphetamine derivatives, such as motor hyperactivity (Halliwell et al., 1964), increased hypothalamic self-stimulation (Stein, 1962) and hyperthermia (Jori and Garattini, 1965). The toxicity of MAO inhibitors, too, is increased (Loveness and Maxwell, 1965). However, tricyclic antidepressants also inhibit the hydroxylation and therefore the inactivation of amphetamines (Sulser et al., 1966, Valzelli et al., 1967, Consolo et al., 1967). It is therefore questionable whether potentiation of amphetamines by tricyclic antidepressants is effected (exclusively) via a direct influence on the NA metabolism.

Another fact indicative of potentiation of the central NA effect is that tricyclic antidepressants antagonize the hypokinesia provoked by substances which deplete MA stores, e.g., reserpine and tetrabenazine (Domenjoz and Theobald, 1959). Animals treated with compounds of this type in fact serve as "depression models" in screening new potential antidepressants.

Inhibition of uptake. Which mechanism underlies this potentiation? Tricyclic antidepressants do not increase the intracerebral MA concentration (Sulser et al., 1962). It is plausible that NA potentiation by tricyclic antidepressants is based on blocking of the (re-)uptake of NA into NA-ergic neurons. The principal arguments in favor of this theory will now be listed.

Imipramine reduces the uptake of intravenously injected ^{3}H-labeled NA in several peripheral tissues, and induces a substantial increase of the plasma level of ^{3}H-labeled NA in the first 5 minutes after the injection (Axelrod et al., 1961). The uptake of circulating NA is dependent on intactness of sympathetic nerve endings (Hertting et al., 1961), and consequently it must be in this tissue that imipramine interferes with NA uptake. Imipramine and other tricyclic compounds also inhibit the uptake of ^{3}H-labeled NA into brain slices (Dengles and Titus, 1961; Ross and Renyi, 1966). Using the histochemical fluorescence technique evolved by Hillarp and co-workers, Hamberger and Masuoka (1965) demonstrated that imipramine and related compounds inhibit the uptake of NA into central and peripheral CA-ergic neurons. However, numerous other compounds were found to have the same effect; they include the neuroleptic chlorpromazine (Dengles and Titus, 1961; Axelrod et al., 1961).

Since NA cannot pass the blood-brain barrier, it was initially impossible to study the influence of imipramine on NA uptake in intact brains. Glowinski et al. (1964, 1965a,b, 1966a,b) solved this problem by direct intraventricular injection of tritiated NA of high specific activity in the rat. The ^{3}H NA is taken up into the NA-ergic nerve endings and, in biochemical terms, behaves like the endogenous transmitter. It was demonstrated that imipramine and related compounds such as desmethylimipramine and amitriptyline (but not chlorpromazine) reduced the NA uptake by 25-40%. Carlsson et al. (1966) supplied yet another argument: in rats whose cerebral NA stores were depleted with the aid of reserpine, restoration of the NA concentration after DOPA administration was prevented by protriptyline and, to a lesser degree, by desmethylimipramine. They assumed that l-DOPA and NA make use of the same membrane pump.

Inhibition of uptake can explain why various tricyclic compounds antagonize the hypothermia and sedation which occur when small amounts of NA are introduced into the mouse brain (Brittain, 1966).

Finally, another ingenious argument was supplied by Carlsson et al. (1969b). They used a number of tyramine derivatives which enter the CA-ergic neuron, supersede both NA and DA from the storage vesicles, and thus reduce the CA concentration. To a varying extent, this effect can be prevented with tricyclic antidepressants of several types. The most plausible explanation is that these compounds block the uptake of tyramine derivatives just as they block that of 5-HT and NA. Carlsson et al. regard the ability of a tricyclic antidepressant to protect NA neurons from the store-depleting effect of tyramine derivatives as a measure of its ability to block NA uptake into the neuron. As measured by this yardstick, protriptyline and desmethylimipramine are strong inhibitors of NA uptake, while imipramine and amitriptyline are feeble inhibitors.

The depletion of DA stores by tyramine derivatives is not antagonized by tricyclic compounds. This indicates that the DA uptake mechanism is not inhibited by antidepressants of this type.

NA degradation. Changes in the intracerebral NA metabolism can be used as a third type of argument in favor of uptake inhibition. After pre-treatment with imipramine or related compounds, the pattern of degradation of intraventricularly injected tritiated NA changes. Methylated products increase, while deaminated products decrease (Schildkraut et al., 1969, Glowinski et al., 1965a,b, Schanberg et al., 1967). Deamination of NA is probably an intraneuronal process, whereas methylation is extraneuronal. These data, too, indicate the probability that penetration of NA into the neuron is impeded by tricyclic compounds.

However, this phenomenon can also be explained differently. To begin with, it is possible that tricyclic antidepressants have a blocking effect on *intra*neuronal membranes as well, specifically those around mitochondria, and thus block the interaction of NA with mitochondrial MAO. One argument in favor of this theory is that, in animals pre-treated with tricyclic antidepressants, reserpine causes no sedation but, instead, motor hyperactivity. This phenomenon is probably based on an increased availability of CA at the postsynaptic receptors, for no excitation is induced when the animals are pre-treated in addition with α-methylmetatyrosine—a substance which depletes CA stores (Sulser et al., 1964). It is conceivable that the excitation is based on the following course of events. Reserpine blocks the uptake of NA into the stores. Tricyclic antidepressants block the mitochondrial membrane to NA so that degradation by MAO fails to occur. NA "leaks" into the synaptic cleft and excites NA receptors. However, there is no direct evidence of this effect of tricyclic antidepressants.

Another possibility, of course, is that tricyclic antidepressants inhibit MAO. It is generally assumed that they do not (Sulser et al., 1964). However, this view needs revision. These compounds do indeed inhibit MAO, *in vitro* (Gabay and Valcourt, 1968; Halaris et al., 1973, Roth and Gillis, 1974) as well as *in vivo* (Edwards and Burns, 1974). MAO has proved to be not a homogeneous system but a complex of isoenzymes. Yang and Neff (1973) distinguished two types: type A MAO, which chiefly regulates the degradation of 5-HT and NA, and type B MAO, which deaminates β-phenylethylamine (PEA). Roth and Gillis (1974) recently published the interesting report that imipramine inhibits in particular type B MAO. This is of interest because a deficiency in physiological PEA is assumed to play a role in the pathogenesis of depressions (Mosnaim et al., 1973).

Tricyclic Antidepressants and 5-HT Uptake

Potentiation of peripheral and central 5-HT effects. The influence of tricyclic antidepressants on the peripheral effects of 5-HT is not unequivocal. Some 5-HT effects are potentiated, e.g., contraction of the feline nictitating membrane (Sigg et al., 1963), whereas other effects are antagonized, e.g., contraction of the canine urinary bladder (Gyermek et al., 1960). It is nevertheless certain that tricyclic antidepressants are able to inhibit the uptake of 5-HT into certain cells of the organism, e.g., blood platelets (Stacey, 1961, Tuomisto, 1974). I mention these platelets by way of example because these cells are regarded as a valid model of serotonergic nerve endings (Pletscher, 1968) and can, moreover, be obtained from human subjects without difficulty. In patients treated with tricyclic antidepressants, the 5-HT uptake capacity of the blood platelets is diminished (Murphy et al., 1970).

Less is known about the function of the central serotonergic than about the central CA-ergic systems. There are nevertheless some central 5-HT effects which are potentiated by tricyclic antidepressants, e.g., the hyperthermia provoked in rabbits by 5-HTP (Loew and Teaschler, 1966). And there are indications that in human subjects, too, certain mental effects of 5-HTP are potentiated by tricyclic compounds (Van Praag et al., 1974).

Inhibition of uptake. Like that of NA, the uptake of 5-HT in the brain is probably inhibited by tricyclic antidepressants. Let us first marshal the *in vitro* arguments. The uptake of 5-HT into brain slices and isolated nerve endings is inhibited (Blackburn et al., 1967; Segawa and Kurumae, 1968). Admittedly, it is uncertain whether we may indeed extrapolate from the synaptosome fraction obtained by differential centrifugation of brain to MA-ergic nerve endings, but it seems probable for it is possible to selectively destroy 5-HT endings in the brain by causing a lesion in the raphe nuclei: the source of the corresponding cell bodies. The synaptosome fraction of brains so treated still takes up NA, but hardly any 5-HT (Kuhar et al., 1972).

Another *in vitro* argument: Rabbits were treated with a MAO inhibitor and then with a tricyclic antidepressant. The cerebral cortex was cut into slices which were incubated in a Krebs–Henseleit's solution. The slices were found to release 5-HT to the medium; the ability to retain 5-HT had been reduced. Carlsson et al. (1969c) explained this phenomenon as follows: 5-HT, which, due to MAO inhibition, is not degraded, "leaks" from the neurons, fails to return to them because of blocking of the 5-HT pump, and appears in the incubation medium.

An important *in vivo* argument is that the uptake into the brain of 5-HT introduced into the lateral ventricle is inhibited by tricyclic compounds (Carlsson et al., 1968). Carlsson et al. (1969a) also used another approach. They observed that certain tyramine derivatives reduce the central 5-HT and NA concentrations, probably because they are taken up into the synaptic vesicles, where they supersede the corresponding amines. Several tricyclic antidepressants antagonize 5-HT depletion in response to these tyramine derivatives. Carlsson et al. regarded this as an indication of a blocking of the 5-HT pump, their basic assumption being that the tyramine derivatives are dependent, for their entry into the neuron, on the same membrane pump as 5-HT.

Differential Influence of Tricyclic Antidepressants on 5-HT and NA Uptake

In all test arrangements, the various tricyclic antidepressants differ in their influence on the uptake of 5-HT and that of NA. The secondary amines, specifically protriptyline and desmethylimipramine, are potent inhibitors of NA uptake but exert relatively little influence on 5-HT uptake. The reverse was found to apply to such tertiary amines as imipramine, amitriptyline and particularly chlorimipramine, the most potent inhibitor of the 5-HT pump in the tricyclic series (Carlsson et al., 1969a, b, c; Tuomisto, 1974). This also applies to 5-HT uptake into human blood platelets (Todrick and Tait, 1969).

As already pointed out, the DA uptake mechanism is probably not inhibited by tricyclic antidepressants (Carlsson et al., 1969b).

Tricyclic Antidepressants and 5-HT and NA Turnover

Tricyclic antidepressants reduce the turnover rate of 5-HT and NA in the brain (Corrodi and Fuxe, 1968, Schubert, 1973). It has been demonstrated with the probenecid technique that this phenomenon probably occurs in human subjects also: the response of 5-HIAA in the CSF to probenecid diminishes during medication with tricyclic compounds (Bowers, 1974, Post and Goodwin, 1974). It is possible that uptake inhibition and turnover reduction are related in the sense that increased availability of 5-HT and NA at the postsynaptic receptors reduces the firing rate in the presynaptic element via a negative feedback mechanism, as a result of which the transmitter production is reduced. The firing rate of serotonergic neurons (i.e., the number of impulses which they generate per unit of time) is indeed reduced by tricyclic antidepressants (Sheard et al., 1972). A reduced firing rate might explain the fact that the intracerebral 5-HT concentration is not decreased in response to tricyclic antidepressants, although the rate of 5-HT synthesis diminishes. However, the reduced turnover can also be explained differently. There are *in vitro* indications that imipramine inhibits the uptake into synaptosomes, not only of 5-HT but also of the 5-HT precursor, tryptophan. A similar effect *in vivo* might prompt reduction of the 5-HT turnover (Bruinvels, 1972).

Conclusions

Numerous findings support the hypothesis that tricyclic antidepressants inhibit the uptake of NA and 5-HT into NA-ergic and serotonergic neurons. The theory that they centrally block the *re*-uptake of 5-HT and NA into these neurons is derived from these findings but is not based on *direct* arguments. Tricyclic antidepressants should not be lumped together as substances influencing the central MA, for they differ widely in the extent to which they influence the NA and the 5-HT pump. There are no unequivocal arguments in favor of a correlation of the biochemical and therapeutic effects listed. The scanty data available will be discussed in Chapter XII.

Iprindole is an exception in the series of tricyclic antidepressants in that it influences neither the uptake nor the turnover of NA in the rat brain (Rosloff and Davis, 1974). For the time being, this fact does not upset the conventional hypothesis on the mechanism of action of tricyclic antidepressants. To begin with, not enough is known about the influence of iprindole on 5-HT (re-) uptake. Secondly, it is possible that the human central MA metablism is indeed influenced; this has not been investigated. Finally, the literature presenting convincing evidence of the antidepressant action of iprindole is scanty.

CHAPTER IV
MAO Inhibitors and Central Monoamine Metabolism

MAO inhibitors are assumed to increase the concentration of MA in the synaptic cleft. Via excitation of postsynaptic MA-susceptible receptors, this is believed to increase the activity in MA-ergic systems. The mechanism responsible for this effect is believed to be MAO inhibition. Inhibition of this enzyme leads to reduced intraneuronal degradation of MA. Consequently, their concentration in the cytoplasm should increase, and "leakage" of MA into the synaptic cleft should occur. The net effect of this process is increased because some MAO inhibitors also inhibit the uptake of MA into the neuron (Hendley and Snyder, 1968). The arguments on which this theory is based are briefly presented in the following sections.

MAO Inhibitors and MA Metabolism

Increased MA concentration. All MAO inhibitors which have been used in the treatment of depressions (a) inhibit the enzyme MAO (Zeller et al., 1959); (b) block the degradation of MA in the brain (Pletscher et al., 1960, 1966); (c) cause an increased concentration of these compounds in the brain (Bogdanski et al., 1956); (d) enhance the increase in MA concentration which results from administration of MA precursors such as l-DOPA and 1-5-HTP (Pletscher et al., 1960). MAO inhibition and increased MA concentration have also been established in the human brain (MacLean et al., 1965; Bevan-Jones et al., 1972).

The increase in MA concentration occurs only at sites where the degradation of MA largely depends on MAO (Pletscher, 1968), and this indicates that the increased MA concentration is indeed a result of MAO inhibition. Another indication in this direction is the fact that the CA degradation pattern changes (Kopin and Axelrod, 1963). The concentration of deaminated CA metabolites (formed in response to MAO) decreases, while that of methylated CA metabolites increases.

In human individuals, the renal excretion of 3-methoxy-4-hydroxymandelic acid (VMA) and 5-HIAA—deamination products of NA and 5-HT, respectively—decreases in response to MAO inhibitors (Van Praag, 1962; Schildkraut et al., 1964). The accumulation of 5-HIAA and HVA in the CSF in response to probenecid likewise decreases, and this indicates that in the human brain, too, the production of deamination products diminishes (Kupfer and Bowers, 1972).

Increased extracellular MA concentration. There are a few indirect arguments that MAO inhibitors do increase the extracellular MA concentration. To begin with, this process has been visualized with the aid of histochemical fluorescence techniques (Fuxe et al., 1966). It is plausible, moreover, because the concentration of methylated CA metabolites increases in response to MAO inhibitors, and the enzyme responsible for this methylation (COMT) is assumed to be localized extraneuronally (Kopin, 1964). As regards 5-HT, in isolated perfused ganglia MAO inhibitors prompt an increase in the 5-HT concentration of the perfusion fluid; this implies that the amount of 5-HT in the extracellular space has increased (Gertner et al., 1957). A final argument is found in the fact that the MA turnover diminishes in response to MAO inhibitors (Masuoka et al., 1963; Glowinski et al., 1972). This phenomenon could be a result of feedback inhibition, initiated by postsynaptic receptor stimulation.

MAO inhibition and antidepressant effect. MAO inhibitors also inhibit several enzymes other than MAO. It is nevertheless plausible that the antidepressant effect relates in particular to MAO inhibition. MAO inhibitors are, in terms of chemical structure, a heterogeneous group of compounds whose common characteristic is their ability to inhibit MAO. Conversely, compounds closely related to MAO inhibitors in chemical structure, but unable to inhibit MAO, are ineffective in depressions (Van Praag, 1962). In human subjects, moreover, a positive correlation has been demonstrated between the degree of MAO inhibition and the degree of clinical improvement (Dunlop et al., 1965). Finally, it has been established that the MA concentration in the human brain gradually increases during medication with MAO inhibitors and attains a peak concentration after about two weeks—at which time the therapeutic effect usually also becomes manifest (MacLean et al., 1965; Bevan-Jones et al., 1972). It is, of course, a logical inference that inhibition of uptake also plays a role in the therapeutic effect of MAO inhibitors. According to Hendley and Snyder (1968), the correlation between therapeutic effect and uptake inhibition is in fact closer than that between therapeutic effect and MAO inhibition. This statement, however, is based on more or less arbitrary (ill-tested) ranking of MAO inhibitors according to therapeutic efficacy.

A correlation between MAO inhibition and antidepressant effect does not necessarily mean that the latter effect is based on an increased concentration of transmitter MA. As pointed out, MAO inhibitors reduce the rate of CA synthesis, and as a result, at least in principle, more tyrosine remains available for the synthesis of other amines such as octopamine and tyramine. Their function is still unknown, but it cannot be excluded that they may contribute to the therapeutic effect.

MAO vs. MAO's

I have so far used the terms MAO and MAO inhibitors without qualification. This has proved to be an unacceptable generalization. There are several types of MAO which differ in substrate specificity, i.e., their readiness to oxidize a given amine and their susceptibility to various MAO inhibitors. The group of MAO inhibitors, therefore, is not a homogeneous group in terms of their influence on the enzyme MAO (Sandler and Youdim, 1972; Neff and Yang, 1974).

Neff and Goridis (1972) made a gross distinction between type A and type B MAO inhibitors. Substrates of predilection for type A are 5-HT and NA; the substrate for type B is β-phenylethylamine. DA and tryptamine are good substrates for both enzyme types. Type A can be more or less selectively inhibited by clorgyline, and type B by deprenyl.

Both types of MAO are present in the human brain (Neff and Yang, 1974). There is no evidence of their being spatially separated, of type A having a predilection, say, for NA and 5-HT neurons. However, the plausibility of a distribution of tasks has been demonstrated (Yang and Neff, 1974). A selective type A inhibitor increases the intracerebral concentrations of 5-HT, NA and DA. A selective type B inhibitor does not influence 5-HT and NA concentrations, but increases the DA concentration.

Virtually all MAO inhibitors so far used in the treatment of depressions are able to block both types of MAO. They are therefore nonspecific MAO inhibitors. It is quite possible that the development of MAO inhibitors with a "focused" effect may yield good antidepressants, with fewer side effects than their "broad-spectrum" predecessors.

Conclusions

It is an established fact that MAO inhibitors increase the central MA concentration in human individuals. This effect is possibly based not only on MAO inhibition but also on inhibition of uptake. In any case, MAO inhibition seems to be a mechanism essential to their therapeutic effect, but the relative significance of the accumulating amines is still obscure. Since MAO have been identified as a group of related enzymes which differ in substrate specificity, and since efforts are being made to evolve selective MAO inhibitors, further elucidation of this problem can be expected.

CHAPTER V

Reserpine: Mirror Image of the Antidepressants

Reserpine and Depression

For some people, reserpine is a depressogenic substance. This was established in hypertensive patients, in some 15% of whom it provokes depressions of the vital type (Goodwin, 1972). I am deliberately using the term "provokes" rather than "causes," because it is uncertain whether causation or provocation is involved. The latter seems the more probable, because the reserpine depression occurs predominantly, if by no means exclusively, in patients with a history of depression. The phenomenon is dose-dependent and occurs mainly at daily reserpine dosages of 0.75 mg or more. The onset of depression is usually observed in the first to seventh month after institution of this treatment. Discontinuation of reserpine as a rule does not lead to abatement of symptoms, nor is this to be expected. Vital depressions can be provoked by a wide variety of factors—psychological, social or somatic—but once they are manifest they take an autonomic course, quite independent of the provoking factors.

Reserpine and Central MA Metabolism

Reserpine has proved to be a mirror image of the antidepressants, not only in psychopathological but also in biochemical terms; unlike the antidepressants, it reduces the amount of MA available at the central receptors.

In 1955, Shore et al. demonstrated that reserpine drastically reduces the 5-HT concentration in the brain. Shortly after, the same was established for the CA (Carlsson et al., 1957). Reserpine inhibits the uptake into and/or storage in the synaptic vesicles of MA. The latter's forced continued presence in the cytoplasm leads to degradation by MAO and a decrease in their concentration. The transmitter is degraded intraneuronally and is deprived of a chance to exert a physiological influence. The MA-ergic activity diminishes. Reserpine does not interfere with the passage of MA through the neuronal membrane; in this respect, it differs from the tricyclic antidepressants, which inhibit this transport process and probably have no effect on intraneuronal storage.

Numerous observations support this conception of reserpine action. Isolated synaptic vesicles from peripheral sympathetic neurons are able to take up and concentrate MA. This process is blocked by reserpine (Stjärne, 1964). Studies of the subcellular distribution of tritiated NA with the aid of differential centrifugation point in the same direction. In animals pre-treated with reserpine, the tritiated NA concentration both in the particle fraction and in the supernatant was decreased. If MAO is inactivated, reserpine inhibits the uptake of tritiated NA into the particle fraction to the same extent, but this time a substantial amount of labeled NA accumulates in the supernatant (Stitzel and Lundborg, 1967). A final argument lies in the fact that, in response to reserpine, the concentration of deaminated CA metabolites increases, while that of methylated CA metabolites decreases. Since deamination is largely an intraneuronal process and methylation an extraneuronal one, this is indicative of an increased intraneuronal degradation of CA (Glowinski et al., 1966a,b).

Reserpine Sedation in Animals and MA Depletion

In test animals, reserpine induces hypomotility but leaves the waking state intact. Is this related to intracerebral MA deficiency? The answer is probably yes, for this effect is antagonized by substances which increase the MA concentration at the central receptors (e.g., tricyclic antidepressants and MAO inhibitors), provided MA synthesis is undisturbed (Manara et al., 1967; Scheckel and Boff, 1964; Sulser et al., 1964; Stein and Ray, 1960). However, this relationship is not a simple one. The degree of sedation does not correspond with the severity of MA depletion, and the two phenomena do not coincide in time: 48 hours after reserpine administration, behavior has returned to normal but the MA concentrations are still decreased. Lundborg (1963) demonstrated in rabbits that, after reserpine administration, the ability to store NA is recovered before the total tissue concentration of NA is normalized. This has led to the hypothesis that restoration of the behavior pattern is determined by restoration of a small but functionally active pool of amines which can vary in size without causing manifest measurable changes in the total amine concentration (Haggendal and Lindqvist, 1964).

Reserpine influences the three transmitter amines to the same degree and within the same period of time. Which of the three has a key function in producing the reserpine sedation? α-MT, an inhibitor of CA synthesis, causes sedation (Weissman et al., 1966). This is in favor of the importance of the CA. Of the two CA with a transmitter function, DA is the more likely candidate. To begin with, l-DOPA antagonizes the reserpine syndrome (Carlsson et al., 1957) and this CA precursor is converted chiefly to DA in the brain. However, this argument is less strong than it may seem to be because, although l-DOPA hardly increases the NA concentration, it does activate the central NA turnover (Keller et al., 1974). Another argument in favor of DA is that dihydroxyphenylserine does not abolish reserpine sedation (Creveling et al., 1968). This compound is directly decarboxylated to NA, without intermediate DA synthesis, and therefore increases the NA concentration selectively.

The role which 5-HT plays in this respect is obscure. pCPA, an inhibitor of 5-HT synthesis, induces no sedation (Koe and Weissman, 1966), but of course it may be that the 5-HT deficiency so achieved is inadequate for this purpose. 5-HTP, a 5-HT precursor which at least in part is converted to 5-HT in central 5-HT neurons (Korf et al., 1974), does not abolish the reserpine syndrome (Carlsson et al., 1957). This would seem to be a conclusive argument against 5-HT as a factor in reserpine sedation, but it is not. 5-HTP by itself can sedate test animals (Modigh, 1973). This is probably a result of peripheral 5-HT influences, for when the peripheral conversion of 5-HTP to 5-HT is inhibited (with the aid of a peripheral decarboxylase inhibitor), 5-HTP behaves like an activating substance. It is unknown whether 5-HTP in combination with a peripheral decarboxylase inhibitor is able to arrest the reserpine syndrome.

Reserpine Depression and MA Depletion

Nothing is known about the influence of reserpine on the central MA metabolism in man, nor about any possible biochemical differences between reserpine-treated patients who do and who do not develop depressive symptoms. ECT is often effective against reserpine depressions (Goodwin, 1972), and this is to be expected in view of the predominantly vital character of these depressions. Antidepressants have not systematically been tested in these cases, but in view of our personal observations there is no reason to doubt their efficacy in this respect.

A correlation between MA depletion and reserpine depression would be probable if abolition of the MA deficiency would cause the depression to disappear. We know of one study in which l-DOPA was successfully administered in reserpine depressions (Degkwitz et al., 1960). 5-HTP has not been tried. Another argument in favor of the above-mentioned correlation is that tetrabenazine, a compound with an influence on central MA which resembles that of reserpine, also seems to be capable of provoking depressions (Lingjaerde, 1963).

Conclusions

Reserpine reduces the effective concentration of MA at the central receptors. This effect is probably responsible for the sedative effect of reserpine in test animals. It is uncertain whether one particular MA plays a key role in this respect, or, if so, which one. A correlation is also suspected between human reserpine depression and MA depletion. Arguments in favor of this theory are virtually all derived from animal experiments. There has been hardly any systematic search for arguments in human patients.

CHAPTER VI
Serotonin Metabolism in Affective Disorders
Study of Peripheral Indicators

Urinary Excretion of Indoleamines

5-HIAA is the principal metabolite of 5-HT. The amount excreted in urine is a gross index of the overall turnover of 5-HT. This is why the urinary 5-HIAA excretion has been determined in depressions, particularly in the early years of MA research in depressions. It was, in fact, the first strategy used to test the MA hypothesis. Several authors found the 24-hour excretion diminished (Pare and Sandler, 1959; Van Praag and Leijnse, 1963a,b), but normal values have also been reported (Cazzullo et al., 1966) and some authors found increased values (Tissot, 1962). Longitudinal studies showed that the level of excretion during manic phases exceeded that during depressive phases (Ström-Olsen and Weil-Malherbe, 1958). Moreover, it was believed that patients with a low 5-HIAA excretion showed a more favorable response to MAO inhibitors than those with higher excretion values (Pare and Sandler, 1959; Van Praag and Leijnse, 1963a,b).

Little is known about the renal excretion of 5-HT. The tryptamine excretion can be either increased (McNamee et al., 1972) or decreased (Coppen et al., 1965b) in depressions.

I do not think that too much value should be attached to observations of this type. Most of the 5-HIAA (and tryptamine) in the urine originates from the periphery, and it is unlikely that the peripheral 5-HT turnover could be indicative of the 5-HT turnover in the CNS. Inversely, it is very questionable whether changes in central 5-HT turnover (given an unchanged peripheral turnover) would lead to measurable changes in renal 5-HIAA excretion. In many of these studies, moreover, the diet was not standardized, and it was later found that 5-HT occurs in certain foods, particularly fruits (Crout and Sjoerdsma, 1959). This may have influenced the results.

5-HT Synthesis

Coppen et al. (1965a) made the interesting observation that, in depressive patients, the production of (^{14}C) carbon dioxide (in expiratory air) from intravenously injected ^{14}C-labeled 5-HTP is decreased as compared with that in a control group. This might suggest a defect in the conversion of 5-HTP to 5-HT. However, these investigators were unable to reproduce their results (Coppen, 1967).

There has since been no duplication of this study, but a variant was carried out. Copen et al. (1974) administered ^{14}C-labeled l-tryptophan to 5 severely depressed patients, and measured the urinary excretion of (^{14}C) 5-HIAA in the depressive phase and after recovery. They found no significant differences between pre-therapeutic and post-therapeutic values, nor between patients and controls. The results showed marked interindividual and intraindividual variations. These observations showed that an overall disorder in 5-HT synthesis in depressed patients is very unlikely. A similar indication is found in the fact that the urinary 5-HIAA excretion after oral administration of tryptophan to depressive patients is not subnormal (Cazzullo et al., 1966). However, Coppen et al. (1974) did demonstrate a slightly decreased tryptophan tolerance in depressions. They measured the blood total tryptophan level at different intervals after intravenous tryptophan loading and found that, during the first 3 hours after loading, the tryptophan concentrations in depressive patients were slightly higher than those in the controls. This might indicate a retardation of the chemical "processing" of tryptophan, in whatever direction, but it might also be that in depressive patients tryptophan is distributed over a smaller volume.

All in all, there is no reason at this time to assume that in depressions there is a defect in the 5-hydroxylation and decarboxylation of tryptophan in the organism in its totality. This need not apply to the CNS. It is to be borne in mind, moreover, that the above-mentioned studies involved only a small number of patients, and that on the other hand it has been demonstrated in recent years that disorders of MA metabolism in depressive patients are not a universal phenomenon but occur in certain subcategories which are sometimes, but not always, identifiable on psychopathological grounds. Studies in small groups carry the risk of missing such biochemical disorders.

Availability of Tryptophan

Only a small amount of tryptophan is used for the synthesis of the indole derivatives 5-HT and tryptamine. A much larger amount is converted via kinurenin and xanthurenic acid to the B-vitamin nicotinic acid, or utilized in protein synthesis (Fig. 10). Two types of observation suggest that the amount of tryptophan available for 5-HT synthesis is possibly diminished in depressions.

Activation of the kinurenin route. Depressive patients convert more tryptophan via the kinurenin route than normal subjects. This was concluded from the fact that in these patients the urinary xanthurenic acid excretion is increased, both basally and after tryptophan loading (Cazzullo et al., 1966). Curzon (1969) reported the same in women with endogenous depressions. Moreover, in two patients suffering from manic-depressive changes who were intravenously injected with radioactive tryptophan, the excretion of radioactive kinurenin was higher during the depressive than during the manic and the normal phases (Rubin, 1967). Why should an abnormally large amount of tryptophan be converted via the kinurenin route? A possible explanation is a high level of circulating hydrocortisone (cortisol), which is a common phenomenon in depressions (Hullin et al., 1967). Corticosteroids from the adrenal cortex stimulate the synthesis of tryptophan pyrrolase in the liver (Knox and Auerbach, 1955); tryptophan pyrrolase is the first enzyme in the conversion of tryptophan via the kinurenin route. In corroboration of this, Rubin (1967) found that an increased plasma cortisol level and increased kinurenin excretion go hand in hand.

A cardinal question in this context is, of course, whether activation of tryptophan pyrrolase does indeed withdraw a substantial amount of tryptophan from 5-HT synthesis. This is plausible. In the acute experiment, hydrocortisone induces in rats (a) activation of liver pyrrolase, (b) a decrease in plasma tryptophan, and (c) a decrease by about 30% of the cerebral 5-HT and 5-HIAA concentrations (Curzon and Green, 1968). When the animals are given a pyrrolase inhibitor (e.g., allopurinol or yohimbine) in advance, hydrocortisone no longer causes a decrease in intracerebral 5-HT (Green and Curzon, 1968). α-Methyltryptophan—a compound which like hydrocortisone activates liver pyrrolase, but independent of the adrenal cortex—likewise causes a decrease in central 5-HT concentration (Curzon, 1969). It is therefore justifiable to conclude that an increased plasma concentration of adrenocortical hormones causes a decrease in intracerebral 5-HT via an increase of the pyrrolase activity in the liver. An observation of interest in this respect was reported by Mangoni (1974), who found that imipramine and the MAO inhibitor tranylcypromine (both antidepressants) inhibit the tryptophan pyrrolase activity in the rat liver.

Plasma tryptophan concentration. According to Coppen et al. (1973), the plasma concentration of free tryptophan in women with "depressive illness" (with no manic phases in the history) was decreased as compared with that in a matched control group. The total tryptophan concentration was not decreased. This implies that the tryptophan fraction bound to plasma proteins was increased at the expense of the unbound (free) tryptophan. After clinical recovery,

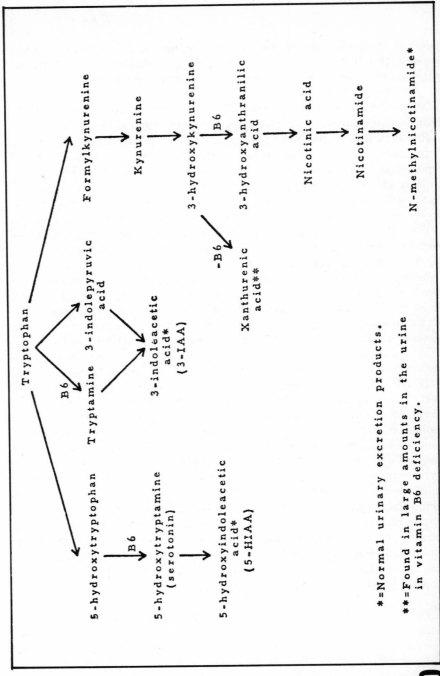

Tryptophan

5-hydroxytryptophan → B6 → 5-hydroxytryptamine (serotonin) → 5-hydroxyindoleacetic acid* (5-HIAA)

B6 → Tryptamine → 3-indoleacetic acid* (3-IAA)

3-indolepyruvic acid → 3-indoleacetic acid* (3-IAA)

Formylkynurenine → Kynurenine → 3-hydroxykynurenine

3-hydroxykynurenine → B6 → 3-hydroxyanthranilic acid → Nicotinic acid → Nicotinamide → N-methylnicotinamide*

3-hydroxykynurenine → -B6 → Xanthurenic acid**

*=Normal urinary excretion products.

**=Found in large amounts in the urine in vitamin B6 deficiency.

10

50

the free tryptophan concentration showed a significant increase without attaining total normalization. Rees et al. (1974) studied two patients with rapidly alternating manic and depressive phases and found a decreased plasma tryptophan concentration in the depressive, and an increased level in the manic phases. They determined only total tryptophan. In some patients with slower alternation of phases, this phenomenon was less pronounced.

Coppen's findings seem important, but need verification and corroboration. The food intake was neither standardized nor controlled, and this may have influenced the results. The test subjects were not receiving drugs that change the degree of tryptophan binding (e.g., salicylates). However, there are other potential interfering factors. Free fatty acids in the plasma, for example, influence the protein-binding of tryptophan (Curzon et al., 1973; Lipsett et al., 1973), and this fat fraction responds strongly to psychological stress; it is therefore possible that the decreased free tryptophan concentration in these cases was a nonspecific phenomenon in relation to the depression.

Conclusions

Studies of the periphery have so far failed to produce convincing indications of a specific disturbance in 5-HT metabolism in depressions. They have revealed a number of suspicious phenomena: a decreased concentration of free tryptophan in plasma, and indications that an abnormally large amount of tryptophan enters the kinurenin "channel." Perhaps the plasma tryptophan concentration is decreased because its removal via the kinurenin route is increased. It is possible that these phenomena correspond more with the degree of anxiety than with the depressive affect as such.

In view of the available data, an overall defect in the conversion of tryptophan to 5-HT cannot be excluded with absolute certainty, and it would be worthwhile to continue investigations along these lines.

FIGURE 10

Tryptophan is involved in protein synthesis. Moreover, it is the mother substance of 5-HT, tryptamine and the B-vitamin nicotinic acid.

CHAPTER VII

Metabolism of Catecholamines in Affective Disorders

Study of Peripheral Indicators

CA and Some of Their Metabolites in Urine and Blood

The renal excretion of NA and normetanephrine can be decreased in depressive patients (Greenspan et al., 1969; Bunney et al., 1970). The patients served as their own controls in these studies. The term *decreased* therefore denotes "lower than during the symptom-free periods." In other words, the NA and normetanephrine concentrations proved to increase when the depression subsided (Schildkraut et al., 1966). Normetanephrine is a methylated NA metabolite. Since the 0-methylation of CA is assumed to occur mostly outside the neuron, normetanephrine is regarded as an indicator of NA-ergic activity (Axelrod, 1966). Decreased NA and normetanephrine excretion is not a universal phenomenon in depressions. It is believed to occur predominantly in retarded (vital) depressions (Schildkraut, 1975).

Is this phenomenon consistent with the hypothesis that depressions can involve a central NA deficiency? I believe so, but it is a marginal argument. To begin with, NA and NA metabolites in the urine are in only a small proportion derived from the CNS. They chiefly reflect the production of adrenaline and NA in peripheral sympathetic nerves and in the adrenal medulla. Moreover, there is no reason to assume that peripheral NA-ergic activity is representative of the central activity. Secondly, the CA excretion is influenced by muscular activity (Karki, 1956). It is therefore possible that the decreased excretion of NA and normetanephrine is related to reduced motor activity. Takahashi et al. (1968) measured activity, but were unable to demonstrate this. Nevertheless, the possibility strikes me as quite real, not only because it is difficult to measure reliably the "production" of motor activity in human subjects, but also because the excretion of NA and normetanephrine has proved to be increased in manic syndromes (Greenspan et al., 1969) and in agitated depressions (Sloane et al., 1966)—syndromes accompanied by motor hyperactivity. Also, the plasma NA concentration can be increased in depressions, and this phenomenon was found not to be correlated with the severity of the depression but rather with the degree of anxiety. It is to be noted, however, that the degree of anxiety and the motor status of these patients were not separately assessed (Wyatt et al., 1971). Finally, the human DA metabolism correlates closely with the patient's motor status (Van Praag et al., 1975b).

All these findings indicate the necessity of an open mind to the possibility that decreased excretion of NA and normetanephrine correlates with reduced motor activity. In a number of manic-depressive patients with rapid phase alternation, Bunney et al. (1970) found that the NA excretion increased on the day preceding the change to the manic phase. This supplies an argument against an exclusive motor determinant of the decreased NA excretion.

The quantitatively most important NA metabolite in the periphery is 3-methyoxy-4-hydroxymandelic acid (VMA). No author has consistently found a decreased excretion of this compound in depressions. In manic patients, Campanini et al. (1970) reported an increased excretion. Greenspan et al. (1969), however, measured normal values even in patients with an increased NA and normetanephrine excretion.

The excretion of DA generally follows the pattern of NA excretion: an increase in manic patients, and a tendency to decrease in depression (Messiha et al., 1970; Schildkraut, 1975).

Urinary MHPG Excretion: A Peripheral Indicator of Central NA Activity?

The principal NA metabolite in the CNS is not VMA but the glycol MHPG, which in the periphery is a subordinate metabolite (Schanberg et al., 1968). Generally speaking, some 50% of the urinary MHPG should originate from the brain (Maas et al., 1973b). The urinary MHPG is therefore regarded as a (gross) indicator of central NA degradation. A low excretion is considered to be indicative of a decreased release of NA from presynaptic neurons (e.g., as a result of a decreased firing rate or an increased re-uptake of NA into the neuron); an increased excretion is regarded as an indication of an increased release of NA (e.g., due to increased activity in the system or a diminished susceptibility of postsynaptic NA receptors, which is compensated for by increased release) (Schild-kraut, 1975).

In depressive patients, the MHPG excretion can be decreased in comparison with the excretion measured in the absence of depressive symptoms. The phenomenon has also been observed in agitated depressive patients with an increased excretion of NA and normetanephrine. It is consequently unlikely that the decreased MHPG excretion relates exclusively to the patient's motor status (Greenspan et al., 1970; Maas et al., 1968). In two manic-depressive patients, the MHPG excretion during manic phases exceeded that during depressive phases, the values obtained during the symptom-free intervals being intermediate (Bond et al., 1972). The MHPG excretion already started to rise during the days preceding the manic change. This, too, indicates the likelihood that MHPG does not primarily reflect the patient's motor status. But this does not imply that MHPG excretion is necessarily related to the mood level. It has been demonstrated, for example, that MHPG excretion increases in response to various types of stress (Maas et al., 1971; Rubin et al., 1970). The question as to which components of the depressive and the manic syndrome, respectively, the MHPG excretion correlates with remains moot.

A decrease in MHPG excretion is not a universal phenomenon in depressions. Schildkraut (1974), studying a small group of patients, found a lower excretion in the manic-depressive subgroup (n = 5) than in patients with chronic characterogenically determined depressions (n = 5). A study on a larger scale was carried out by Maas et al. (1973a): patients with so-called "primary" depressions (n = 21) had a 24-hour excretion of 1032 ± 63 μg MHPG, vs. 1348 ± 65 μg in the control group. The difference was highly significant (p = 0.0005). Primary depressions are predominantly vital depressions, which occur in individuals with a relatively undisturbed personality structure. In terms of clinical course, they may be manic-depressive (= bipolar) syndromes, unipolar depressions, and vital depressions occurring for the first time. In the study of Maas et al., the bipolar subgroup showed the lowest excretion values, but the difference from the other subgroups was not significant. This observation, however, was confirmed in a subsequent study (Deleon-Jones et al., 1975).

Tyrosine in Blood

In patients with vital depressions, the fasting plasma tyrosine values do not deviate from the normal (Takahashi et al., 1968). Decreased values have been measured in the morning hours (0800-1100 h) (Birkmayer and Linauer, 1970; Benkert et al., 1971). Of greater interest, it seems to me, are the observations of Takahashi that patients with bipolar depressions attain higher plasma tyrosine values after oral administration of tyrosine than do controls. Benkert et al.

(1971), however, found an entirely normal tyrosine "tolerance" in similar patients.

Enzyme Studies

Apart from the excretion pattern of CA metabolites in depressions, studies have been devoted to enzymes involved in the CA metabolism.

Catechol-O-methyltransferase (COMT). NA which is released upon excitation of NA-ergic nerves is largely eliminated, after transmission of the information, by re-uptake into the neuron. A small amount is extraneuronally degraded, and in this process COMT plays an important role. This enzyme specifically methylates catechol groups. It is found throughout the organism, but circulating NA is O-methylated mainly in the liver (Hertting and La Brosse, 1962). The fact that COMT also occurs in the erythrocytes has facilitated its study in human subjects (Axelrod and Cohn, 1971).

Two studies (Cohn et al., 1970; Dunner et al., 1971) disclosed about 50% less COMT activity in the erythrocytes in women with unipolar depressions than in a control group. In women with bipolar depressions, the COMT activity was likewise decreased, but less markedly so. Women under other diagnostic headings, including schizophrenia, showed no abnormalities of COMT activity. Surprisingly, the decreased COMT activity was found to be definitely sex-linked. It has not been found in men with unipolar or bipolar depressions.

The importance of this phenomenon is obscure. It has so far been observed exclusively in depressions, but it is not syndrome-linked: the COMT activity was also found decreased when the depressive symptoms had subsided. In depressive patients, there are more disorders of the MA metabolism which persist after clinical recovery. It is possible that these are factors predisposing to the manifestation of depressions (see chapter X).

It is unknown whether the decreased peripheral COMT activity is indicative of the situation in the brain. Assuming for the moment that it is, this would support the hypothesis advanced by Prange et al. (1970): that the susceptibility of the central CA receptors is diminished in depressions. There are indeed some peripheral indications to this effect. For example, the influence of infused NA on the systolic blood pressure is diminished in depressive patients (Prange et al., 1967). A decreased COMT activity can be conceived of as a compensatory mechanism, providing an offset against the diminished receptor susceptibility. This, however, is still a purely hypothetical construction.

Monoamine oxidase (MAO). The data on this enzyme, which plays an important role in MA degradation, are controversial. In women with severe "endogenous" depressions (all premenopausal), a plasma MAO activity was measured which was far higher than (six times as high as) the normal value (Klaiber et al., 1972). No information was given on the occurrence or nonoccurrence of (hypo)manic phases. Now, it is a fact that peripheral as well as central MAO activity increases with increasing age (Robinson et al., 1972), but age differences could not explain the difference between the depressive and the control group. Oral administration of conjugated estrogens led to normalization of the MAO activity and abatement of the depression. In this study, however, there was no control group. The authors concluded that their findings are consistent with the hypothesis that a CA deficiency (possibly, therefore, as a result of MAO overactivity) plays a role in the pathogenesis of depressions. The findings of Sandler et al. (1975) point in the same direction. They found a deficit in tyramine conjugation in a group of depressed patients. Most of the amine was degraded via oxidative deamination. They postulated that the conjugation deficit was secondary to increased MAO activity.

Murphy and Weiss (1972), however, found a 50% decrease in MAO activity in depressions of the bipolar type. A corresponding finding was an increased renal excretion of tryptamine—an amine whose degradation is entirely dependent on MAO. These authors measured the MAO activity, not in plasma but in blood platelets. In the four patients in whom this question was studied, the MAO activity continued to be decreased also after subsidence of the symptoms, and also in the (hypo)manic phase. In patients with depressions of the unipolar type, the average MAO activity was slightly (10%) but not significantly increased. The MAO activity of blood platelets in women exceeds that in men. This applies to all age categories (Robinson et al., 1971). However, since the three groups studied were matched as to sex and age distribution, this factor cannot have influenced the findings obtained.

Gottfries et al. (1974) recently found that the MAO activity in various areas of the brain in suicide victims (n = 15) was 20—40% lower than that in a comparable control group (n = 20) with a different cause of death. The decrease was demonstrable regardless of whether tryptamine of β-phenylethylamine was used as substrate for MAO. However, it seems questionable whether these observations indeed corroborate those reported by Murphy and Weiss. It is unknown whether the patients in the suicide group had all been suffering from depressions, let alone whether these depressions were of the bipolar type.

Is a decreased central MAO activity, if any, inconsistent with the CA deficiency hypothesis? It need not necessarily be. In the first place, the phenomenon could be a secondary one—some sort of compensatory mechanism aimed at abolition of the assumed CA deficiency. In that case, however, it would hardly have any physiological significance, because a 20-40% reduction of MAO activity is not likely to be sufficient to retard CA degradation. Another possibilty would be that, as a result of primarily decreased MAO activity, a nontransmitter amine such as octopamine would accumulate in CA-ergic neurons. Such a compound could function as a false transmitter, and reduce or arrest the activity in the corresponding circuit. An argument in favor of such a possibility is the fact that octopamine accumulates in blood platelets following administration of MAO inhibitors, and that the same has been observed in depressive patients with an (endogenously) decreased MAO activity in the blood platelets (Murphy, 1972).

It is not clear why the data on MAO are so controversial. Perhaps the characteristics of MAO in plasma differ from those of MAO in blood platelets. It has been known for some years now that MAO comprises a complex of enzymes which have different properties, and that the enzyme's composition can vary from one organ to the next. Moreover, MAO activity can be temporarily influenced by hormonal influences (e.g. estrogens) and—at least in animals—by certain states of behavior (Klaiber et al. 1971; Eleftherion and Boehlke, 1967). In human subjects, these factors are not entirely controllable, and may have influenced results.

Neuroendocrine Factors as "Markers" of CA-ergic Activity

If intracerebral MA-ergic pathways in depressive patients are indeed hypofunctional, disturbances are to be expected in the systems regulated by these pathways. Conversely, a disorder in such a system can be validly viewed as an argument that something is amiss with the central MA-ergic pathways. Today there are numerous data which indicate the likelihood of a correlation between biogenic amines in the brain and neuroendocrine regulation (Anton-Tay and Wurtman, 1971). From this point of view, too, neuroendocrinological research in depressive patients certainly seems to have significance, and it is surprising that to date so little has been done.

The secretion of prolactin by the anterior hypophysis is regulated by DA-ergic neurons of the tubero-infundibular DA system. The secretion is inhibited by intensification of DA-ergic activity, and stimulated by its diminution. This is believed to take place via an increase and decrease, respectively, of the secretion of PIF (prolactin-inhibiting factor)—a hormone-like substance formed by neuroendocrine cells in the hypothalamus which inhibits hypophyseal prolactin secretion (Meites et al., 1972).

The growth hormone shows the reverse of this behavior: its secretion increases with increased central CA-ergic activity and decreases with its decrease (Martin, 1973). This regulation is probably effected via an increase and decrease, respectively, of the secretion of GRF (growth hormone-releasing factor) by neuroendocrine cells in the hypothalamus. There are indications that NA (Toivola and Gale, 1972) as well as DA (Müller et al., 1970) are involved in the regulation of growth hormone secretion.

l-DOPA—precursor of DA as well as of NA—causes a decrease in the human plasma prolactin concentration, and an increase in the growth hormone concentration (Kleinberg et al., 1971; Kansal et al., 1972). The growth hormone level also rises in response to insulin-induced hypoglycemia; this is probably a CA-ergic effect, for it disappears following administration of reserpine (depletion of

CA stores) and α-MT (inhibition of CA synthesis) (Miller et al., 1967).

The prolactin response to l-DOPA administration was found to be normal in patients with unipolar and bipolar depressions (Sachar et al., 1973), and so was the growth hormone response (Gruen et al., 1975). The growth hormone response to insulin-induced hypoglycemia, however, was lower in women with unipolar depressions (all in the menopause) than in a control group (Gruen et al., 1975). The authors consider this finding to be consistent with the hypothesis that a functional intracerebral NA deficiency exists in certain types of depression. Then why is it that the DOPA response was normal? Possibly (but this is speculation) because the DOPA response is produced via the DA system and the hypoglycemia response via the NA system, the former being normal while the latter is hypofunctional.

Cyclic AMP and Central CA-ergic Activity

Cyclic AMP probably plays a role in transmission in MA-ergic neurons. The relation to the CA-ergic system is best known. In the effector cell, the transmitter stimulates the synthesis of cyclic AMP, and it is this so-called "second messenger" that evokes the physiological response in the effector cell. This is why, in the context of the MA hypothesis, a decreased concentration of cyclic AMP is to be expected in the brain in depressive patients.

In fact, the urinary excretion of cyclic AMP was found to be decreased in depressive phases and increased in manic phases (Abdulla and Hamadah, 1970; Paul et al., 1970, 1971). The phenomenon has been demonstrated both in vital and in personal depressions (Sinanan et al., 1975). It is believed to be related to mood in that the most severely depressive patients also showed the most marked decrease in excretion (Paul et al., 1971). Moreover, the excretion of cyclic AMP was found to increase in response to ECT (Hamadah et al., 1972) and antidepressant medication (Ramsden, 1970)—an additional argument in favor of its relation to mood. Others, however, maintained that the changed excretion of cyclic AMP relates more to the motor status: retardation in depression and unrest in mania (Berg and Glinsmann, 1970; Eccleston et al., 1970). Cramer et al., (1972) reported a decreased post-probenecid concentration of cyclic AMP in the CSF in depressive patients, but Robison et al. (1970) found a normal concentration.

The significance of these findings, however, is questionable. Cyclic AMP is found in such wide distribution throughout the organism that the renal excretion of this compound can hardly be expected to give information on its central utilization. Moreover, there are indications that changes in the cyclic AMP level in the central CA-ergic systems are not reflected in the concentration of cyclic AMP in the CSF (Sebens and Korf, 1975).

For the time being, it does not look as if the cyclic AMP concentrations in CSF and urine are a marker of central CA-ergic activity.

Conclusions

(1) The pattern of excretion of NA and NA metabolites provides some indications that overall NA-ergic activity can be decreased in (vital and primary) depressions, and increased in manic phases. However, these phenomena are probably related not so much to mood changes as to "nonspecific" phenomena such as motor activity and degree of anxiety. Generally speaking, biological depression research is inclined to directly relate biological variables to the specific symptom (in this case the mood changes), and to pay little attention to the possible influence of the above-mentioned nonspecific symptoms. It seems quite possible that this can lead the investigator astray. The scanty data on tyrosine metabolism in depressions are not unequivocal, and therefore not interpretable.

(2) Interesting attempts have been made to test the hypothesis of a CA deficiency indirectly, i.e., by the functioning of CA-ergically regulated mechanisms. The growth hormone response to insulin-induced hypoglycemia was found decreased in women with unipolar depressions. Since the hypophyseal growth hormone secretion is believed to be CA-ergically regulated, this can be regarded as an indication that a functional intracerebral CA deficiency in fact exists. Other investigators have measured concentrations in CSF and urine of cyclic AMP, a "second messenger" interposed between CA release and the physiological responses of the effector cell. The findings are contradictory, and this is not surprising because cyclic AMP is far from exclusively related to the CA-ergic system, but instead plays a role in numerous cells throughout the organism.

(3) In women with unipolar depressions, the COMT activity in the erythrocytes is decreased. This phenomenon also occurs in bipolar depressions, but is less pronounced. It is not observed in men. The decrease persists after clinical recovery and therefore seems to be a predisposing factor *(a locus minoris resistentiae)* rather than a causative factor.

The data on MAO are not unequivocal. There are reports on decreased as well as increased peripheral MAO activity in depressions. One study reports a decreased MAO activity in the brain in suicide victims. However, it is not known whether all these victims had been suffering from depressions. These observations are incompatible unless we resort to an artifice and assume that investigators who found increased and those who found decreased MAO values happened to study depressive patients in different phases of the disease process.

CHAPTER VIII

Post-mortem Study of the Central MA Metabolism in Depressions

It is conceivable that the CNS can supply more information on the central MA metabolism than the periphery. The problem is how to obtain this information in human subjects. This has been tried in three different ways: post-mortem examination of the brains of suicide victims, determination of the concentration of MA metabolites in the CSF (the mother substances do not occur in the CSF in measurable amounts), and determination of the accumulation of acid MA metabolites in the CSF after probenecid loading. The results thus obtained will be discussed in this and the next two chapters.

Indoleamines

The first study of this type was published in 1967, when Shaw et al. reported that the 5-HT concentration in the hindbrain was lower in a group of suicide victims than in a control group of individuals who had died from accidents or from acute somatic non-neurological diseases. The 5-HIAA concentration was not measured. Pare et al. (1969) likewise found a decreased 5-HT concentration, but reported normal values for 5-HIAA. In the study of Bourne et al (1968), no differences in 5-HT concentration were found, but the 5-HIAA concentration was found decreased in the suicide group. The last study published on this subject was of great importance (Lloyd et al., 1974): the brain stem was studied not in its entirety but in separate components. A significant decrease in 5-HT concentration was found exclusively in the raphe nuclei—not in all nuclei but only in the dorsal and the inferior central nucleus. In both nuclei, the 5-HIAA concentration was lower than that in the control group, but the difference was not statistically significant. It may be mentioned in passing that the system of raphe nuclei is the site of predilection of 5-HT in the CNS.

One post mortem study remained negative; it is that by Gottfries et al. (1976), who did not separately examine the raphé nuclei. Moreover only one out of 23 post mortem cases were considered to have suffered from endogenous depression.

Catecholamines

In two of the four post-mortem studies mentioned above, the concentration of CA was also measured. The NA concentration in hindbrain (Bourne et al., 1968) and hypothalamus (Pare et al., 1969) showed no significant differences between suicide group and control group, nor did the DA concentration in the caudate nucleus (Pare et al., 1969).

Monoamine Oxidase

Gottfries et al. (1975) compared MAO activity in the brains of 15 suicides of whom 8 were alcoholics to a control material of 20 individuals without known mental disorders. MAO activity was determined in 13 different parts of the brain with β-phenylethylamine and tryptamine as substrates.

The MAO activity in all parts of the brain investigated was found to be significantly lower in the alcoholic suicides as compared to controls, while there was no significant difference between the nonalcoholic suicides and controls.

Apparently there exists a connection between low MAO activity in the brain and suicidal behavior among alcoholics.

Significance of Post-mortem Findings

Chemical data obtained after death are delicate material in terms of interpretation. They can have been influenced by a variety of uncontrolled or uncontrollable factors such as use of drugs prior to death (several of the test subjects had committed suicide with the aid of drugs), death causes in the control group, interval between death and post-mortem examination, influences of age and sex, the duration of the final agony, etc. On the other hand, doubts about the method as such cannot obliterate the fact that the results of four out of five studies all point in the same direction: that of a deficiency of 5-HT and 5-HIAA in the suicide group. This situation might be based on a diminished 5-HT turnover.

That the results of these studies show a fair degree of agreement is surprising particularly from a psychiatric point of view. Suicide is not a diagnosis but the fateful terminal point of all sorts of developments: overwhelming problems of life, a depression in the psychiatric sense, addiction, a psychotic state, etc. Information on the histories of the suicide victims involved was frequently inadequate. It seems unlikely that they had all been suffering from pathological depressions. The biochemical data, therefore, would seem to be more homogeneous than the psychopathological features. An explanation of this paradox might lie in the fact that disorders of 5-HT metabolism are related less to the clinical entity known as depression than to a decrease in the mood level, regardless of the syndrome within which this occurs, and its etiology. This would be in agreement with my view that the biochemical dysfunctions found in the brains of psychiatric patients in the past few years can be better and more sensibly correlated with disturbances in separate mental functions than with syndromes, i.e., complex patterns of disturbed behavior, or nosological entities (see chapter X).

Conclusions

Post-mortem findings indicated that the 5-HT and 5-HIAA concentrations in the lower portion of the brain stem in suicide victims were decreased. MAO activity was found to be lowered in the brains of alcoholic suicides, a finding which is of course difficult to reconcile with decreased levels of 5-HT. The CA concentration did not differ from that in a control group. The decrease in indoleamines might indicate that the 5-HT turnover had been low prior to death. Since a group of suicide victims is probably not very homogeneous in psychopathological terms, the suspicion arises that the disorders of central 5-HT metabolism observed were related less to the syndrome depression, or a given nosological concept of depression, than to a lowered mood level, regardless of the syndrome within which this phenomenon occurs, or its etiology.

CHAPTER IX

Central MA Metabolism in Affective Disorders

Cerebrospinal Fluid Studies Without Probenecid

MA Metabolites in CSF: Do They Reflect the Central MA Metabolism?

The concentration of MA metabolites is usually measured in lumbar CSF. Do these values give a more or less reliable impression of the metabolism of the mother amines in the CNS? The following arguments indicate that indeed they do.

In the CNS itself, there is of course a relation between the concentration of a MA metabolite in a given area, and the amount of amine locally metabolized. Animal experiments have demonstrated a correlation between the 5-HIAA and HVA concentrations in the CSF, and their concentrations in adjacent parts of the CNS (Moir et al., 1970; Papeschi et al., 1971; Eccleston et al., 1968). Under normal conditions, moreover, there is no transport of any significance of amine metabolites from the periphery to the CNS. This has been demonstrated for 5-HIAA and HVA in test animals (Bartholini et al., 1966; Bulat and Zivković, 1971) and for (conjugated) MHPG in human individuals (Chase et al., 1973).

The principal argument that the human lumbar CSF can supply relevant information on central MA metabolism is that, in response to factors which influence the central MA turnover, the CSF concentration of MA metabolites changes in the expected direction. The 5-HT precursors tryptophan and 5-HTP increase 5-HT synthesis. After administration to human subjects, the 5-HIAA concentration in the CSF increases in accordance with expectation (Dunner and

Goodwin, 1972; Van Praag et al., 1973; Van Praag, 1975). Moreover, there is a significant correlation between tryptophan and 5-HIAA concentrations (Ashcroft et al., 1973), and this is an additional argument that the 5-HIAA concentration in the CSF and the 5-HT turnover in the CNS are linked. After l-DOPA administration, the concentration of HVA in the CSF increases (Goodwin et al., 1970). Exogenous l-DOPA is converted for the most part to DA, only a small amount being converted to NA. The only exogenous method to cause a selective increase in NA synthesis is administration of DOPS (dihydroxy-phenyl-serine)— a compound which is directly converted to NA without intermediate DA production (Creveling et al., 1968). To my knowledge, there have been no efforts to establish whether DOPS does indeed increase the concentration of MHPG in human CSF.

Reduction of MA synthesis also has its reflection in the CSF. In response to pCPA—an inhibitor of tryptophan hydroxylase and therefore of 5-HT synthesis— the 5-HIAA concentration in lumbar CSF decreases (Chase, 1972). The same happens with MHPG in response to α-MT, an inhibitor of tyrosine hydroxylase, and fusaric acid, an inhibitor of DA-β-hydroxylase (Chase et al., 1973). In Parkinson's disease, the DA concentration in the basal ganglia and the HVA concentration in the CSF both show a marked decrease (Bernheimer et al., 1966).

MA Metabolites in Lumbar CSF: Do They Reflect the Cerebral or the Spinal MA Metabolism?

As already pointed out, there are strong arguments in favor of a relation between the concentration of MA metabolites in lumbar CSF and the metabolism of the mother amines in the CNS in human individuals. Another question is whether it is indeed the cerebral MA metabolism that is reflected in the lumbar CSF. Could it be that mainly the spinal MA metabolism is measured in lumbar CSF? Even if this were true, such determinations would not be without value. If depressions involve a defect in the MA metabolism, there is no reason to believe in advance that this disorder would suddenly cease at the level of the spinal cord. On the other hand, the value of determinations in CSF would increase if it could be shown that these probably also give an impression of the cerebral metabolism.

HVA. The most informative test in the latter sense is determination of HVA in lumbar CSF. HVA is really largely of cerebral origin, which is not surprising in view of the fact that, so far as we know, few or no DA-ergic neurons or nerve endings are contained in the spinal cord (Carlsson et al., 1964). Accordingly, the gradient between ventricular and lumbar HVA concentration is high (10 : 1) (Sourkes, 1973). There are other strong arguments in favor of the view that lumbar HVA is representative of the cerebral DA metabolism. Partial block (Curzon et al., 1971; Garelis and Sourkes, 1973; Post et al., 1973) and total block of the spinal subarachnoidal space (Young et al., 1973) leads to a decrease of HVA in (respectively, its total disappearance from) the CSF caudal to the block. After occlusion of both interventricular foramina, only very little HVA was demonstr-

able in the lumbar CSF (Garelis et al., 1974). This demonstrates that HVA largely originates from structures in the vicinity of the lateral ventricles. These findings give substance to the view that HVA in the human and animal lumbar CSF
(Papeschi et al., 1971) originates from DA-rich areas close to the lateral ventricles, mainly from the caudate nucleus (Garelis et al., 1974). Accoring to Sourkes
(1973), an important amount of the HVA there produced is drained into the lateral ventricles (not into the bloodstream). From the lateral ventricles it reaches
the lumbar CSF partly by diffusion and partly in the flow of CSF. Under more
or less basal conditions, the time required for this is estimated to be 4 hours
(Korf and Van Praag, 1970; Pletscher et al., 1967).

While under normal conditions the spinal cord hardly contributes to the HVA
concentration in lumbar CSF, this changes when large doses of l-DOPA are administered. In the rat, at least, this leads to substantial DA synthesis in the spinal
cord (Andén et al., 1972). It seems plausible that in human subjects, too, the
spinal cord contributes substantially to the HVA in lumbar CSF in this case
(Garelis et al. 1974).

5-HIAA. 5-HIAA in lumbar CSF is of mixed spinal/cerebral origin. It could
hardly be otherwise, for serotonergic nerve endings occur in the spinal cord. Accordingly, there is a ventriculo-lumbar gradient of 5-HIAA concentrations but
one which is less high than that of HVA concentrations (10 : 3) (Sourkes, 1973).
It is none the less plausible that some of the lumbar 5-HIAA is of cerebral origin.
The first argument is that, after oral administration of tryptophan to patients,
the tryptophan concentrations in lumbar CSF begins to rise after 2 hours,
whereas that of 5-HIAA does not begin to rise until after 6 hours. The 5-HIAA
accumulation is believed to be delayed because some of the 5-HIAA must first
be transported from the brain to the lumbar CSF (Eccleston et al., 1970). When
transport of 5-HIAA from the CNS is inhibited with the aid of probenecid, the
rise of the 5-HIAA concentration in lumbar CSF also occurs only after a few
hours (Korf and Van Praag, 1970). However, a different interpretation of this
phenomenon is possible, namely, that spinal 5-HT has lower rates of synthesis
and degradation than cerebral 5-HT.

Another argument lies in the fact that in children with noncommunicating
hydrocephalus the ventriculo-lumbar 5-HIAA gradient is higher than that in hydrocephalic children in whom the CSF compartments do communicate (Anders-
sson and Roos, 1969), so that ventricular 5-HIAA has free access to the spinal
subarachnoidal space. The value of this arguments should not be overrated. In
noncommunicating hydrocephalus, the CSF pressure above the obstruction is
increased. In dogs, an increase in CSF pressure was found to delay the 5-HIAA
transport from the CSF (Andersson and Roos, 1968). It is therefore possible
that differences between children with communicating and those with noncommunicating hydrocephalus are explained by differences in hydrostatic pressure.

A final argument can be derived from studies of patients with transverse cord lesions or CSF block. In the former patients the 5-HIAA concentration in lumbar CSF is not decreased (Post et al., 1973). In test animals, severance of the spinal cord leads to about 80% reduction of the 5-HT concentration (Carlsson et al., 1964; Shibuya and Andersson, 1968). If this applies to human subjects also, then the above-mentioned observation is indicative of a substantial contribution of higher levels to the lumbar 5-HIAA level. Given a total block high in the spinal subarachnoidal space, one would expect a decrease in 5-HIAA concentration caudal to the block if a substantial amount of the lumbar 5-HIAA is of cerebral origin, and little change in this concentration if it largely originates from the spinal cord. The relevant data are controversial: Curzon et al. (1971) found decreased 5-HIAA values, but Garelis and Sourkes (1973) and Young et al. (1973) reported normal values. The discrepancy may be based on differences in the state of the cord below the block. It is quite conceivable that production and drainage of CSF at this level could be disturbed in individually different degrees (Garelis et al., 1974).

One argument seemed to be an unqualified refutation of cerebral participation in the 5-HIAA concentration in lumbar CSF. It caused some misgivings in the ranks of investigators of the biological substrates of depressions, because part of their argument that cerebral 5-HT plays a role in the pathogenesis of depressions derived from studies of lumbar CSF. Bulat and Zivković (1971) submitted cats to intracisternal injection of 5-HIAA, and found no increase of 5-HIAA in lumbar CSF. Their conclusion was that all lumbar 5-HIAA is of spinal origin—a rather hasty conclusion, because it is quite possible that the 5-HIAA injected is effectively eliminated from the CSF and disappears before it can reach the lumbar level. Weir et al. (1973) repeated the experiment in a slightly modified form, and obtained different results. A solution containing labeled 5-HIAA was infused into the basal cisterns of the cat at a constant rate. Calculations on the basis of the amount of 5-HIAA and the amount of label in the lumbar CSF demonstrated that a certain amount of the lumbar 5-HIAA (40-70% of its total amount) originated from structures rostral to the foramen magnum.

MHPG. Much less is known about the origin of MHPG in lumbar CSF. In man, there is no gradient between ventricular and lumbar CSF concentrations (Chase et al., 1973), and it is therefore probable that a substantial amount of lumbar MHPG originates from the spinal cord. An argument which points in the same direction is the fact that, in patients with a transverse lesion (whose spinal NA concentration can be assumed to have markedly decreased), the lumbar MHPG concentration is decreased, whereas this is not found in patients with a high-level spinal CSF block (in whom MHPG from rostral areas cannot reach the spinal subarachnoidal space) (Post et al., 1973).

Shortcomings of the CSF Strategy

Determination of the concentration of MA metabolites in lumbar CSF (a procedure which I shall henceforth call "CSF determination" for the sake of brevity) affords information on the metabolism of the mother amines in the CNS, but the method shows some serious shortcomings.

To begin with, it gives an impression of the gross degradation of an amine at a given moment, but not of the amount of amine degraded within a given unit of time (i.e., not of the turnover). Another disadvantage is that the MA metabolites in lumbar CSF are not all of the same origin. MHPG is of spinal, 5-HIAA of mixed spinal/cerebral, and HVA of cerebral origin. In actual cases, the information obtained by CSF determination is rather heterogeneous. The last disadvantage I would mention is the most serious flaw: the method is not very reliable. This is true for several reasons. To begin with, MA metabolites are not altogether transported to the bloodstream via the CSF. A certain amount enters the bloodstream directly; for 5-HIAA, this amount is estimated to be no less than 90% (Meek and Neff, 1973). For other metabolites, this fraction is unknown. As such, a substantial transport of MA metabolites directly to the bloodstream does not devaluate CSF determination, provided the ratio between directly and indirectly transported material is constant; yet it is precisely on this point that we have no information whatever.

Another factor detrimental to the reliability of CSF determination is the fact that the flow rate of the CSF is subject to the influence of CSF pressure (Davson, 1967). The amount of a MA metabolite encountered in the lumbar CSF at a given moment is in part determined by the CSF flow. This means that the concentration is affected by a variety of factors which influence CSF pressure, e.g., postural changes, straining, sneezing, coughing—all factors which it is difficult either to eliminate or to normalize.

The above-mentioned factors probably explain the substantial day-to-day differences in the CSF concentration of MA metabolites in a given individual (Van Praag et al., unpublished data); they also explain why, as we shall see, the findings in depressive patients are not unequivocal.

The shortcomings of CSF determination can be largely avoided by premedication with probenecid. This procedure has consequently improved the study of the central MA metabolism in living human individuals to a considerable extent.

MA Metabolites in the CSF in Affective Disorders

Depression. In 1960, Ashcroft and Sharman reported that in depressive patients the CSF concentration of substances with a 5-hydroxyindole structure is decreased. In the course of subsequent years, it was demonstrated that 5-HIAA is involved in these cases, and several investigators corroborated this observation (Ashcroft et al., 1966; Denker et al., 1966; Van Praag et al., 1970; Coppen et al., 1972; Mendels et al., 1972). This phenomenon can indicate a diminished 5-HT metabolism in the CNS. However, not all results are unequivocal. Some authors described the decrease as slight and not significant (Bowers et al., 1969; Papeschi and McClure, 1971); others found no decrease at all (Roos and Sjöström, 1969; Goodwin and Post, 1973; Wilk et al., 1972; Sjöström and Roos, 1972).

The data on the HVA concentration are more homogeneous. The majority of investigators found decreased values in lumbar CSF (Bowers, 1969; Roos and Sjöström, 1969; Papeschi and McClure, 1971; Mendels et al., 1972), although the decrease was not always significant (Van Praag and Korf, 1971; Wilk et al., 1972). The data on MHPG, on the other hand, are confusing. Its concentration has been decribed as decreased (Gordon and Oliver, 1971; Post et al., 1973), but also as normal (Shopsin et al., 1973). Shaw et al. (1973) found a significant decrease in MHPG concentration after recovery from depression, indicating that it had been increased in the depressive phase. In human CSF, MHPG occurs in free form (70%) and as sulfate ester (30%). In the above-mentioned studies, the total MHPG concentration was determined. Whether the ratio of the fractions cited is changed in depressive patients remains unknown. According to Jimerson et al. (1975), not only CSF MHPG but also CSF vanillylmandelic acid (VMA)—a minor NA metabolite in the human brain—is reduced in depressed patients.

Mania. Manic phases have been much less fully studied than depressive phases, probably because manic patients cannot be deprived of medication for several days without difficulty, and do not readily adjust themselves to a strict research protocol. Several investigators reported a decreased lumbar 5-HIAA concentration (Denker et al., 1966; Bowers et al., 1969, Coppen et al., 1972, Mendels et al., 1972). This is the more remarkable because patients of this category often show motor unrest—a factor in response to which the lumbar 5-HIAA concentration often tends to assume higher values, probably due to more effective "mixing" of the CSF (Post et al., 1973). This finding could be a biological argument that mania and depression do not represent a polarity but have the same roots—a view which psychodynamic investigators have advocated on quite different grounds (Mendels et al., 1975).

However, the findings in mania are not consistent, either: normal and increased 5-HIAA values have also been reported (Bowers et al., 1969; Goodwin and Post, 1973, Wilk et al., 1972).

In HVA studies, several but not all (Mendels et al., 1972) investigators found a decreased concentration in the lumbar CSF (Bowers et al., 1969; Roos and Sjöström, 1969; Wilk et al., 1972; Papeschi et al., 1971; Coppen et al., 1972). According to Post et al. (1973), normal or increased concentrations are found in very markedly disinhibited patients who cannot be confined to bed, and in whom the "mixing effect" probably manifests itself.

The MHPG concentration was found to be significantly increased by Wilk et al. (1972) and Shopsin et al. (1973) in manic patients, but Post et al. (1973) found normal values. There have so far been no reports on a decreased MHPG concentration. Interpretation of the increased MHPG values should take into account that free MHPG can pass the blood-brain barrier, and that consequently peripheral MHPG can contribute to the concentration found in lumbar CSF (Goldstein and Gerber, 1963). However, this contribution cannot be very great, for peripheral MHPG is largely conjugated, and in this form it is unable to pass the blood-brain barrier (Shopsin et al., 1973). Accordingly of intravenously injected labeled MHPG only very little proved to appear in the CSF (Post et al., 1973).

Causes of the Variability of the Results of CSF Studies

It can be stated that more or less decreased 5-HIAA and HVA concentrations in lumbar CSF have been fairly regularly found in depressive and manic patients, but these findings have by no means been consistent. There can be several reasons for this.

(1) In many cases, the syndromes studied were not systematically and consistently differentiated by three criteria: symptomatology, etiology and course. Instead, the results in the depression group as a whole were averaged, and compared with those in a control group. This method is apt to conceal possible differences between subgroups, and there are indications that this did happen. We ourselves, for example, regularly observed a decreased lumbar 5-HIAA concentration in the group of vital depressions, but only sporadically in the group of personal depressions (Van Praag et al., 1970, 1973)—findings which were corroborated by Mendels et al. (1972). The criterion "course" is probably also important: in unipolar depressions the decrease is believed to be more pronounced than that in the bipolar group (Mendels et al., 1972; Ashcroft et al., 1973).

According to Van Praag and Korf (1971), the lumbar HVA concentration in retarded depressions tended to show lower values than that in nonretarded depressions and that in a control group. However, the differences were not significant on a 5% level. Other authors found no differences in this respect (Mendels et al., 1972; Wilk et al., 1972). In this context, too, the criterion "course" may be of importance. Ashcroft et al. (1973) found decreased HVA concentrations in unipolar depressions, but not in the bipolar group.

The decreased MHPG values reported by Post et al. (1973) were observed in a relatively large group (n = 25) of patients with primary depressions. No other depression categories were investigated. Shopsin et al. (1973) found no decrease, but their group was small and comprised a mixed bag: 6 "endogenous" and 2 "reactive" depressions.

So far, no correlations have been established between CSF concentrations of 5-HIAA and HVA on the one hand, and severity or etiology of the depressions on the other. All in all, there are indications that the concentrations of MA metabolites in CSF differ in different categories of depression. This is why generalizing statements on the biochemical characteristics of "depression" should be viewed with the necessary reservations.

(2) Differences in age might play a role. According to Bowers and Gerbode (1968), the 5-HIAA concentration in CSF increases with increasing age, and is significantly higher over than under age 60. Ashcroft et al. (1973) found a more complicated relation between age and 5-HIAA, with minimal 5-HIAA values in age group 40-50 and higher values in both younger and older individuals. Other authors have found no age correlation (Goodwin and Post, 1973; Papeschi and McClure, 1971, Nordin et al., 1971).

(3) The control groups generally comprised no normal test subjects but consisted of hospitalized patients with neurological or psychiatric conditions. None showed marked abnormalities of mood, but as a group they were far from homogeneous.

(4) The baseline 5-HIAA concentration proved to show substantial day-to-day variations in the same individual. Possible causes were discussed in a previous subsection. Perhaps this also explains why the subnormal values which we initially found in vital depressions (Van Praag, 1969; Van Praag et al., 1970) were not reproduced in a reduplication study (Van Praag et al., 1973).

(5) It is questionable whether all studies were made under standardized conditions. An important point, for example, is that the patient be confined to bed for a few hours preceding the CSF determination. For 5-HIAA and HVA, after all, there is a ventriculo-lumbar concentration gradient, and motor activity promotes CSF mixing with, as a possible result, an increase in lumbar concentration (Post et al., 1973). To ensure comparable results, moreover, it is necessary always to examine the same CSF sample, e.g., the first 5 ml obtained; as a result of the gradient, the concentration of MA metabolites increases as the CSF originates from a higher level. Finally, it is to be taken into account that food can contain MA, and that the diet must therefore be standardized in this respect (Erspamer, 1966).

Are the CSF Changes Syndrome-Dependent or Syndrome-Independent?

This question has so far received but scanty attention, and therefore no unequivocal answer can as yet be given. Mendels et al. (1972) reported that, after recovery, the HVA concentration increased "substantially," and that of 5-HIAA only slightly. However, not all patients they examined for the second time were free from medication. The findings reported by Coppen et al. (1972) and Ashcroft et al. (1973) are less equivocal: no increase either in 5-HIAA or in HVA.

At the time of the second examination, the patient had been without medication for 5 days. It remains to be established whether this period is sufficiently long. Tricyclic antidepressants, which some of the patients had received, lower the 5-HIAA level, and a continuation of this effect may have concealed possible differences in concentration before and after treatment.

Another objection can be made to both studies. The second CSF determination was made during the hospital period, when the patient was considered to be free from symptoms and a few days after discontinuation of medication. In unipolar and bipolar depressions, the depressive phase has an average duration of 5-6 months (Van Praag, 1962; Angst et al., 1973). Antidepressants bridge the depressive phase but do not arrest it: early discontinuation of medication is often followed by a relapse within one or a few weeks. In the light of this fact, it is questionable whether the above-mentioned patients could justifiably be described as having "recovered."

Shaw et al. (1973) were the only investigators to collect some data on MHPG. In depressive patients, the concentration of this metabolite was not abnormal. After recovery, however, a slight but significant decrease occurred, and this may indicate that the NA turnover had nevertheless been increased in the depressive phase.

Even if further investigation would show that the disorders of the MA metabolism are not syndrome-dependent but persist after recovery, this would not necessarily imply lack of correlation between the two phenomena (Van Praag, 1974). The metabolic disorders might be factors predisposing to the behavior disorders. (chapter X)

Specificity of CSF Findings

A decrease in the 5-HIAA and HVA concentrations in lumbar CSF is not specific for the group of affective disorders, but is observed also in Parkinson's disease (Olsson and Roos, 1968, Bernheimer et al., 1966, Lakke et al., 1972; Puite et al., 1973, Godwin-Austen et al., 1971), in seriously disabled (Claveria et al., 1974) or bedridden multiple sclerosis patients (Sonninen et al., 1973) (the phenomenon is not observed in less serious cases) in motor neuron disease, supranuclear palsy (Claveria et al., 1974), and in presenile and senile dementia (Gottfries and Roos, 1969, 1970).

In Parkinson's disease, the decreased levels are likely to be due to a reduced metabolism of DA and 5-HT, for the phenomenon corresponds with the decreased 5-HT and DA concentrations found post-mortem in the basal ganglia (Ehringer and Hornykiewicz, 1960). In depressions, a metabolic disorder is likewise plausible: first of all because post-mortem examination of suicide victims has revealed a decreased 5-HT concentration in the brain stem, and secondly because the 5-HIAA and HVA concentrations in the CSF increase in response to 5-HT and DA precursors—compounds which promote the synthesis of the mother substances (Goodwin et al., 1970; Dunner and Goodwin, 1972; Van Praag, 1975).

Low CSF 5-HIAA and HVA levels based on a deficient 5-HT and DA metabolism in the CNS are therefore not specific of a given nosological entity or syndrome. But this does not mean that they are nonspecific. They could be *symptomatologically specific,* that is to say, related to particular disorders of motor or psychological function. For example, the decreased DA turnover might be related to motor hypoactivity, and the decreased 5-HT turnover to a pathological lowering of the mood level, regardless of the syndromal and nosological context in which these phenomena occur (Van Praag et al., 1975b).

Of the other syndromes under discussion, it is unknown whether the low 5-HIAA and HVA values are based on metabolic disorders. Post-mortem 5-HT and DA concentrations, and the effects of precursors, are not known. It is also possible that, in chronic degenerative diseases of the CNS such as multiple sclerosis, Pick's disease and Alzheimer's disease, the transport of MA metabolites from the CNS is disrupted. Moreover, immobilization with inadequate CSF mixing can have played a role. An argument in favor of the latter possibility is that low values were only observed in seriously disabled patients with multiple sclerosis. Gottfries and Roos (1970), however, found no evident impoverization of motility in their (pre)senile patients. Moreover, the probenecid-induced accumulation of 5-HIAA and HVA was also reduced and this indicates the likelihood of a decreased metabolsim of the mother amines (Gottfires and Roos, 1973).

In psychiatric syndromes other than the depressive and the manic, few CSF studies have been made. In acute schizophrenia, there have been reports on decreased 5-HIAA values (Bowers et al., 1969, Ashcroft et al., 1966), but also on normal values (Rimon et al., 1971, Van Praag and Korf, 1975). The HVA concentration tends to show a slight decrease (Bowers et al., 1969). This is alleged to be the case particularly in paranoid patients (Rimón et al., 1971). However, the number of patients examined is too small to warrant definite conclusions. So far, the MHPG concentration has not been found to show abnormalities in psychotic reactions (Van Praag and Korf, 1975).

The Ability of the CNS to Synthesize MA

In a group of 10 patients with unspecified depressive syndromes, Coppen et al. (1972) found a decreased tryptophan concentration in the lumbar CSF. The enzyme tryptophan-5-hydroxylase is normally not saturated with tryptophan, and consequently the 5-HT synthesis in the brain varies with the amount of tryptophan available (Jéquier et al., 1967). If the decreased tryptophan concentration in the CSF is representative of the situation in the CNS, this could explain the suspected decrease in 5-HT synthesis. However, Ashcroft et al.

(1973) found normal tryptophan values in unipolar as well as in bipolar depressions, and their findings are in agreement with ours.

The CSF 5-HIAA concentration following tryptophan loading gives an impression of the capacity of the CNS to convert tryptophan to 5-HT. Ashcroft et al. (1973) found no differences in this respect between depressive patients in the unipolar and the bipolar group, and neurological controls. They determined the 5-HIAA concentration 10 hours after tryptophan administration. We carried out a similar experiment, but measured the CSF 5-HIAA and tryptophan concentrations at different times (and, of course, in different patients). A lumbar puncture was carried out after 1, 2, 5 and 8 hours per group of 5 patients. Eccleston et al. (1970) made a similar study of neurological patients without psychiatric changes. In the depression group, the CSF tryptophan concentration was found to be tripled 1 hour after loading, in the nondepressive patients it had hardly changed by that time. Moreover, 8 hours after loading, the CSF 5-HIAA concentration showed a much less marked increase in the depressive patients than in the control group (Van Praag et al., 1973b). More tryptophan in the CSF and less 5-HIAA: this suggests that the tryptophan administered "congests," so to speak, in the CNS in depressive patients, i.e., is less readily converted to 5-HT. This phenomenon could explain a decreased 5-HT turnover.

There are no analogous data on the CA metabolism. The CSF tyrosine concentration in depressive patients is normal (Van Praag et al., 1973). Nor does the conversion of l-DOPA to HVA show any imperfections (Goodwin et al., 1970; Van Praag, 1974).

Conclusions

There are indications that the baseline concentrations of 5-HIAA, HVA and MHPG in the CSF can be decreased in depressions. The indications are most convincing for the group of the vital depressions with a unipolar course, and for the so-called "primary" depressions. It is not clear whether the levels of the three metabolites are interrelated. Similar abnormalities have been found in manic patients (although less widely studied), and this suggests that mania and (vital) depression are pathogenetically related rather than representing the extremes of a range.

The sparse data available point in the direction of a persistence of the biochemical changes, at least of the low 5-HIAA values, after the patient's clinical recovery. Provided it is corroborated, this suggestion permits of two interpretations: (a) the biochemical and behavior changes are unrelated, or (b) the biochemical changes have not a directly causative but a predisposing significance: they render the patient susceptible, as it were, to the occurrence of depressions.

I must stress again that the CSF findings are far from consistent. This can have several causes, but undoubtedly an important one lies in the fact that the method is weak in the sense that it yields results which are not very reliable, and that it provides no more than a vague impression of the MA turnover in the brain and spinal cord. The probenecid technique is much less marred by these imperfections.

CHAPTER X

Central MA Metabolism in Affective Disorders
CSF Studies After Probenecid Loading

Principles of the Probenecid Technique

It was explained in the preceding chapter that the concentration of an MA metabolite in the lumbar CSF reflects the metabolism of the mother amines in the CNS, but that the reflection is a rather blurred one. Its definition can be substantially enhanced by administering probenecid prior to CSF determination.

Probenecid intrarenally inhibits the transport of various organic acids, including 5-HIAA and HVA (Despopoulos and Weissbach, 1957; Werdinius, 1967). It was found that these acids are also eliminated from the CNS by an active transport process, which is inhibited by probenecid (Ashcroft et al., 1968; Neff et al., 1967). Consequently, probenecid administration is followed in all test animals studied by an increase in the concentration of 5-HIAA and HVA in the CNS, both in the brain (Werdinius, 1967) and in the ventricular and cisternal CSF (Guldberg et al., 1966; Bowers and Gerbode, 1968). The increase is a linear one, at least for several hours. This suggests that the degradation of the amines continues undisturbed for some time, and is not immediately reduced by feedback mechanisms.

Assuming that transport from the CNS is totally inhibited, the rate of accumulation of 5-HIAA and HVA should equal the rate of synthesis of these acids; that is to say, it should equal the rate of degradation of the mother amines. Under steady-state conditions, the rates of degradation and synthesis of the mother amines are equal. This means that the rate of accumulation of 5-HIAA and HVA in the CNS following probenecid administration can be accepted as a measure of the turnover of the mother amines. The turnover values measured with the probenecid technique proved, in fact, to correspond well with those obtained by other methods such as the rate of accumulation of an amine after MAO inhibition or the rate of its disappearance after inhibition of synthesis (Neff et al. 1967).

MHPG shows only a slight degree of accumulation following probenecid administration, and is therefore unsuitable for use as a method to study the central NA turnover (Korf et al., 1971; Gordon et al., 1973). The transport system of MHPG apparently has a low sensitivity to probenecid. This applies only to free MHPG, for the transport of MHPG sulphate is certainly probenecid-sensitive (Extein et al., 1973). In the human CNS, however, MHPG is present for the most part in the unconjugated form.

Advantages of the Probenecid Technique Over "Plain"
CSF Determination

The use of probenecid eliminates the principal limitations of "plain" CSF determination. Its advantages can be summarized as follows.

(1) The method is not a "snapshot" of the amine degradation, but gives information on the total metabolite production in a given period of time, i.e., on the turnover of the amine in question.

(2) The method gives an impression of the total metabolite production, whereas "plain" CSF determination gives an impression only of the fraction which leaves the CNS via the CSF since probenecid probably inhibits the direct transport of 5-HIAA and HVA to the bloodstream as well as their indirect transport via the CSF. This can be concluded from the fact that, in animals, 5-HT and DA turnover values measured with the aid of probenecid show good agreement with the values obtained by other methods. The probenecid technique, therefore, eliminates the uncertainty factor which lies in our ignorance of the amounts of 5-HIAA and HVA transported directly, and indirectly via the CSF, to the blood stream.

(3) As a result of the transport block, 5-HIAA and HVA are "caught," so to speak, in the CNS, and the ventriculo-lumbar concentration gradient diminishes. In this way, the lumbar CSF concentration of a metabolite gives a more faithful impression of the degradation of the mother amines in the CNS in its totality. For the same reason, the influence of local changes in CSF pressure on the metabolite concentration in lumbar CSF diminishes. Consequently the intraindividual day-to-day variations in 5-HIAA and HVA concentrations after probenecid are much smaller than those in baseline concentrations.

The post-probenecid accumulation of 5-HIAA and HVA in lumbar CSF thus provides an indication of the turnover of 5-HT and DA in the CNS as a whole. Note the use of the word "indication," as opposed to "measure." The lumbar punctures are generally confined to one before and one after administration of probenecid. This limitation is determined by ethical considerations. The accumulation "curve" so obtained comprises too few points for calculation of the turnover. The probenecid technique would probably provide a true measure of central 5-HT and DA turnover if an acceptable method could be evolved to sample human CSF regularly over a period of say, 12 hours.

Procedure of the Probenecid Test

Unfortunately, the procedure of the probenecid test has not yet been standardized, and the various procedures have yet to be compared. The first studies in human individuals were made with small amounts of probenecid (a total of few grams), distributed over several days (Roos and Sjöström, 1969; Tamarkin et al., 1970, Van Praag et al., 1970; Bowers, 1972). With these amounts, the concentrations of 5-HIAA and HVA in lumbar CSF were approximately doubled. In animal experiments, much larger amounts of probenedid were given, and the increase was many times as large. This indicated that transport inhibition in the human subjects was far from total, and that larger amounts of probenecid would have to be given.

Two dosage schedules are currently in use: that of Goodwin et al. (1973) and that of Van Praag et al. (1973). The former procedure calls for administration of a large dose of probenecid in the course of 18 hours. The first lumbar puncture (LP) is performed at 0900 h, and the second at 1500 h on the following day. In the course of the 18 hours preceding the second LP, the patient receives 100 mg probenecid per kg body weight, divided into four fractional doses (given at 0900, 1400, 0700 and 1200 h). The patient is confined to bed for 8 hours preceding the first LP and throughout the period of probenecid administration. Each LP yields 10 ml CSF, which is collected in 20 mg ascorbic acid and stored at -20° C until determinations are made.

We give a large dose of probenecid (5 g) within a much shorter time (5 h) and partly (1 g) by intravenous drip. This mode of administration was chosen in order to achieve optimal transport inhibition within the shortest possible time, thus minimizing the risk of accumulating metabolites' influencing the metabolism of the mother substances. The following is a detailed description of the procedure we have adopted.

Day 1: The patient is on a standard hospital menu, without fruits containing CA or IA (breakfast at 0800, lunch at 1230 and dinner at 1730 h). He participates in all the usual ward activities, retires at the usual time and remains in bed until 4 hours after the first LP on day 2.

Day 2: The patient is on a standard hospital menu and remains in bed until 4 hours after the LP (at 0830 h, still fasting). Breakfast immediately after the LP. Four hours after the LP, the patient gets up and participates in the usual ward activities until he retires at the usual time.

Day 3: The patient is on a standard hospital menu and participates in the usual ward activities until he retires at the usual time. He remains in bed until 4 hours after the second LP on day 4.

Day 4: The patient is on a standard hospital menu and remains in bed until 4 hours after the LP.

0800 h—breakfast
0830 h—1 g probenecid in 150 ml physiological saline, given by intravenous drip within 20 min

0930 h—1 g probenecid by mouth
1030 h—1 g probenecid by mouth
1130 h—1 g probenecid by mouth
1230 h—1 g probenecid by mouth
1630 h—lumbar puncture

The LP is carried out on the prone patient. The first 10 ml CSF is obtained and collected in one tube. The tube is carefully shaken manually to ensure optimal mixing, for there is a concentration gradient for 5-HIAA and HVA between ventricular, cisternal and lumbar CSF, and the concentrations of these compounds increase in CSF from higher levels.

As soon as it is obtained, the CSF is frozen, and duplicate determinations are made within a week of the following concentrations: 5-HIAA (Korf et al., 1973). HVA (Westerink and Korf 1975) and probenecid (Korf and Van Praag, 1971). This is done because probenecid concentration and 5-HIAA/HVA accumulation are positively correlated (Korf and Van Praag, 1971; Sjöström, 1972): the more probenecid in the CSF, the higher the 5-HIAA and HVA accumulation. This correlation of course ceases to exist when transport inhibition becomes total. It is uncertain whether this stage is actually reached in human subjects, but there are indications that it is. We therefore consider it necessary to have a check as to whether a possible increased or decreased accumulation is in fact due to an increased or decreased production of 5-HIAA and HVA in the CNS, and not to increased or decreased penetration of probenecid into the CNS.

A disadvantage of our procedure is that test subjects not infrequently complain of a heavy sensation in the upper abdomen, nausea and sometimes headaches in the course of the morning (in about 1 out of 3 cases). However, vomiting (which makes the test a failure) occurs only sporadically (in about 1 out of 25 cases). The above-mentioned side effects gradually subside in the course of 3-6 hours after the last probenecid administration.

We have never observed side effects from the intravenous probenecid drip. Intravenous injection of probenecid gives rise to local pain, possibly due to irritation of the vascular wall; this is why this practically simple mode of administration cannot be used (Korf and Van Praag, 1970).

So far, we have made the following observations in all psychiatric (psychoses, depressions) and neurological syndromes (Parkinson's disease) with disturbances in the central 5-HT and DA turnover (Van Praag et al., 1973a; Lakke et al., 1972; Korf et al., 1974). When, after probenecid, the accumulation of 5-HIAA (i.e., the difference between post-probenecid value and baseline concentration) is decreased or increased, the post-probenecid value is likewise decreased or increased, and the levels of significance are virtually the same. This was found also to apply to HVA. The implication is that it is not absolutely necessary to measure *accumulations:* it is sufficient to determine post-probenecid 5-HIAA and HVA concentrations. In the above-mentioned syndromes, we therefore consider it justifiable to perform only one lumbar puncture, particularly when diagnostic rather than research purposes are involved. This is of course a great practical advantage, and makes it easier to obtain permission to repeat the test.

Does the Probenecid Technique Afford Information on Human DA and 5-HT Turnover?

In test animals, there is good agreement between MA turnover calculated with the aid of the probenecid technique, and values obtained by other methods such as determination of the rate of disappearance or accumulation of MA after inhibition of their synthesis or degradation, respectively. This means that the probenecid technique really yields turnover values. In human subjects, this cannot be proven because there are no other methods to measure central MA turnover. The finding that 5-HIAA and HVA accumulation would increase after procedures which increase the turnover of the mother substances, and decrease after procedures with the opposite effect, would be a strong argument that turnover is also measured with the probenecid technique in human individuals. The following subsections show that this argument is indeed valid.

Decrease of 5-HT and DA turnover. Inhibition of the synthesis of 5-HT and DA leads by definition to a decrease of their turnover. If the probenecid technique really measures turnovers, the inhibition of synthesis can be expected to be associated with a decrease in 5-HIAA and HVA accumulation. To begin with, 5-HT and DA synthesis can be inhibited *directly,* with the aid of enzyme inhibitors such as pCPA and α-MT; they inhibit tryptophan and tyrosine hydroxylase, respectively, and these enzymes catalyze the first step in the synthesis of 5-HT and DA, respectively.

There are several indications that the synthesis of 5-HT and DA can also be inhibited *indirectly*, by stimulating postsynaptic 5-HT and DA receptors. Via a feedback mechanism, the firing rate in the presynaptic element is thus reduced, and the synthesis of 5-HT and DA diminishes (Sulser and Sanders-Bush, 1971; Schubert et al., 1970; Sheard et al., 1972; Thoenen, 1972). The 5-HT and DA receptors in turn can be stimulated either directly or indirectly: directly, by means of substances which, like the transmitters themselves, stimulate these receptors, e.g., apomorphine, a DA receptor stimulator (Andén et al., 1967), and quipazine (Grabowska et al., 1974) and LSD (Andén et al., 1968), which stimulate 5-HT receptors; indirectly, by increasing the amount of 5-HT and DA available at the corresponding receptors. This can be done by:

(1) reducing the re-uptake of 5-HT and DA into the neuron (for 5-HT, this can be effected with the aid of tricyclic antidepressants);

(2) reducing the intraneuronal degradation of 5-HT and DA by administration of MAO inhibitors;

(3) facilitating the release of transmitters from the synaptic-vesicles into the synaptic cleft (for 5-HT, this effect can be obtained with certain 4-chloramphetamines).

In human individuals, the synthesis inhibitors pCPA and α-MT suppress the probenecid-induced accumulation of 5-HIAA and HVA almost entirely, and they do this selectively in that pCPA leaves intact the accumulation of HVA, and α-MT that of 5-HIAA (Goodwin et al., 1973).

Of the substances which directly stimulate 5-HT and DA receptors, only LSD has been studied in human subjects. It was found to reduce the 5-HIAA accumulation (Bowers, 1972). More information is available on the indirect stimulators. Two tricyclic compounds have been studied in human individuals: amitriptyline(Bowers, 1972; Post and Goodwin, 1974) and imipramine (Post and Goodwin, 1974). Both compounds reduce the accumulation of 5-HIAA but do not influence that of HVA. It has been found in animal experiments that tricyclic compounds reduce the intracerebral 5-HT turnover (Corrodi and Fuxe, 1969; Schildkraut et al., 1969). As pointed out, this effect is regarded as secondary to inhibition of the 5-HT uptake into the neuron. A direct influence on 5-HT synthesis is not excluded, however, for these substances also inhibit the uptake of the mother substance tryptophan into the neuron (Bruinvels, 1972). Be this as it may, diminution of 5-HT synthesis proves to be associated with decreased 5-HIAA accumulation after probenecid. The unchanged HVA accumulation is consistent with the fact that tricyclic compounds do not influence the (re-)uptake of DA.

Phenelzine, the only MAO inhibitor studied in human subjects, reduces the

HVA accumulation but not that of 5-HIAA (Kupfer and Bowers, 1972). This is against expectation, for phenelzine blocks the degradation of DA and 5-HT and reduces the 5-HT turnover. A possible explanation of this discrepancy lies in the observation of Ashcroft et al. (1969) that phenelzine inhibits the transport of 5-HIAA from the CSF. Assuming that the inhibition of 5-HIAA transport by probenecid has been subtotal, it is conceivable that after phenelzine the 5-HT turnover is more markedly inhibited than the 5-HIAA accumulation in the CSF suggests.

4-Chloromethylamphetamine and 4-chloramphetamine reduce the accumulation of 5-HIAA, but not that of HVA (Van Praag, 1975). In test animals, they reduce the 5-HT turnover (Korf and Van Praag, 1972; Sanders-Bush and Sulser, 1970). In this case, too, decreased 5-HT turnover and decreased 5-HIAA accumulation go hand in hand.

Increase of 5-HT and DA turnover. The synthesis of 5-HT and DA can be directly stimulated by administering the corresponding precursors. There is also an indirect method, at least as far as CA are concerned: blocking the post-synaptic receptors. This leads to an increased firing rate in the presynaptic element, and an increase in transmitter synthesis (Van Rossum, 1967; Thoenen, 1972; Bunney et al., 1973). Via this mechanism, neuroleptics of the phenothiazine and the butyrophenone type are believed to increase the intracerebral turnover of DA and NA (Andén et al., 1972). They do not influence the 5-HT turnover. There are no known compounds which show selective blocking of 5-HT receptors.

The 5-HT precursors tryptophan (Goodwin et al., 1973) and 5-HTP (Van Praag and Korf, 1975d) both cause a substantial increase in 5-HIAA accumulation after probenecid, without exerting a significant influence on that of HVA. The DA precursor l-DOPA causes a very marked increase in HVA accumulation (Goodwin et al., 1973), but it is to be borne in mind that after administration of large doses of l-DOPA some of the peripherally formed HVA penetrates into the CNS; consequently, the HVA concentration in the CSF is no longer a faithful reflection of the DA degradation in the CNS (Prockop et al., 1974). In addition, l-DOPA induces a moderate but significant decrease of the 5-HIAA accumulation. A possible explanation could be that DOPA enters serotonergic neurons and is there converted by the aromatic acid decarboxylase to DA, which in its turn supersedes 5-HT from the stores (Ng et al., 1970).

In human individuals, neuroleptics of the phenothiazine and butyrophenone type cause a dose-dependent increase of HVA accumulation after probenecid; this is indicative of an increased DA turnover (Bowers, 1973; Van Praag and Korf, 1975). They do not influence the 5-HIAA accumulation. No indications of a increased central NA turnover where found (Van Praag and Korf, 1975). It is to be noted, however, that this statement is based exclusively on determination of the baseline MHPG concentration (MHPG transport being probenecid-insensitive), which is less reliable than determination of accumulation after transport block.

The 5-HT and DA turnover and the accumulation of 5-HIAA and HVA in the CSF after probenecid thus prove to be positively correlated. This is a strong argument that the accumulation is an indicator of the turnover of the mother substance.

Shortcomings of the Probenecid Technique

The probenecid technique enables us to make a gross estimate of the 5-HT and DA turnover in the human CNS as a whole, but the technique is by no means flawless. It is unsuitable for the study of the NA metabolism, for example, and for 5-HIAA and HVA it measures the total production in the CNS, which means that relatively small local changes can remain unnoticed.

A second imperfection is uncertainty about the question as to whether 5-HIAA and HVA transport block can be approximately total. There are indications that in fact it is. After administration of 5 g probenecid in 5 hours, the accumulation of 5-HIAA and HVA does not significantly exceed that after 4 g in 5 hours. Moreover, the rising curve shows a plateau at a time when the probenecid concentration in the CSF is still high (Tamarkin et al., 1970; Van Praag et al., 1973). Exact data on the degree of inhibition achieved, however, are not available. In any case, the doses given are the maximum amounts acceptable: more probenecid could not be given. For this reason, the probenecid concentration in the CSF is routinely included in the determinations in our research unit, in order to exclude the possibility that an abnormal accumulation is due to an abnormal probenecid concentration rather than to an abnormal production of metabolites in the CNS.

Another vulnerable feature of this technique is that probenecid also influences the 5-HT metabolism in a manner other than by blocking 5-HIAA transport. In animals (Tagliamonte et al., 1971) and in human individuals (Korf et al., 1972) probenecid reduces the plasma total (free and protein-bound) tryptophan concentration. The plasma free tryptophan concentration increases (Lewander and Sjöström, 1973). Probenecid probably interferes with the binding of tryptophan to serum albumins. This explains the increase of the free fraction. Free tryptophan disappears to the tissues, and consequently the total plasma concentration diminishes. In agreement with this hypothesis, the intracerebral tryptophan concentration increases (Tagliamonte et al., 1971; Korf et al., 1972). The 5-HT turnover rate, however, shows no or hardly any increase (Korf et al., 1972; Barkai et al., 1972). In view of the fact that the intracerebral 5-HT turnover can be reliably measured in animals with the aid of probenecid (Neff et al., 1967), this was to be expected. It thus proves to be possible to vary the intracerebral tryptophan concentration without influencing the rate of 5-HT synthesis. This phenomenon is inconsistent with the theory of Fernström and Wurtman (1971), who maintain that intracerebral tryptophan is an important regulator of the rate of 5-HT synthesis. Probenecid does not influence the human CSF trypto-phan concentration (Korf et al., 1972; Lewander and Sjöström, 1973).

There are therefore no indications that changes in the tryptophan metabolism devaluate the probenecid test.

5-HIAA and HVA are the principal metabolites of 5-HT and DA in the human CNS. There is little doubt about this. However, some synthesis of alcohols such as 5-hydroxytryptophol or 3-methoxy-4-hydroxyphenylethanol cannot be excluded. This would mean that the turnover of 5-HT and DA exceeds the value indicated by the probenecid technique. Finally, it is to be noted in this context that, after probenecid premedication, a small fraction of large doses of intraven-ously given 5-HIAA enters the CNS (Bulat and Zivković, 1973). Since (surpris-ingly) there have been no studies of the influence of probenecid on the peripheral 5-HT metabolism, the possible practical significance of this factor in human individuals cannot be assessed.

The principle of the probenecid test is undoubtedly important: accumulation of a metabolite in the CNS as an index of its rate of synthesis. Probenecid itself, however, is not an ideal transport blocker; there is certainly a need for a better one.

Central 5-HT Turnover in Depressions

Decrease of 5-HT turnover. The probenecid technique, evolved in animal ex-periments by Neff et al. (1967), was first applied to human individuals in Sweden by Roos et al. (1968, 1969), in the Netherlands by Van Praag et al. (1969, 1970) and in the U.S.A. by Bowers (1969) and Goodwin et al. (1970) in virtually simultaneous, but independent investigations.

Toward the end of the Sixties, Roos and Sjöström (1969) and Van Praag et al. (1969, 1970) reported that the 5-HIAA concentration in the human lumbar CSF increases in response to probenecid, and that the increase is less pronounced in depressive patients than in controls. As explained above, a low accumulation of 5-HIAA after probenecid administration is indicative of a low 5-HT turnover in the CNS. Van Praag et al. studied patients with recurrent vital ("endogenous") depressions of varying etiology, as did Roos and Sjöström. Tamarkin et al. (1970) and Bowers (1969) likewise observed 5-HIAA accumulation after probenecid, but they reported no results obtained in depressive patients.

The initial studies were done with relatively small amounts of probenecid. To maximalize transport block, much larger doses were later used. The results thus obtained were similar in principle: an averagely decreased 5-HIAA accumulation in vital depressions of both the unipolar and the bipolar type. In this context, Van Praag et al. (1973a) compared vital depressive with personal depressive syn-dromes, and depressive patients with nondepressive controls. Sjöström and Roos (1972) also studied psychotics and patients with neurotic personality disorders. Both groups of investigators reported that the decrease of accumulation was more pronounced in the bipolar than in the unipolar group. Goodwin et al. (1973) likewise reported lower accumulation values in vital depressive patients than in nondepressive controls. The difference was not significant, but this may have been due to the smallness of the depression group (n = 6).

Negative findings. So far, only Bowers (1972) has reported an averagely normal 5-HIAA accumulation in depressions. He studied 10 patients with depressions of the unipolar type. The literature currently shows an inclination to accept this designation as an adequate qualification of a depression. I do not share this view. The term "unipolar" denotes a clinical course, and gives no information on the nature of the syndrome. Vital depressions often tend to be recurrent, but so do personal depressions, which are often provoked by chronic neurotic personality factors. Moreover, in the confrontation with a first phase it is uncertain whether a depression will take a recurrent course or whether (hypo)manic phases will or will not follow. The term "unipolar" is therefore not an adequate definition of a depression. A syndromal and etiological qualification must be added. When dealing with a first depressive period, moreover, the term is useless. Several of Bowers' patients were described as "anxious" and "hostile." It is therefore quite possible that personal depressions were involved—a category in which the phenomenon in question is hardly ever observed, if at all.

Vital depressions with and without disturbed 5-HT metabolism. We ourselves found indications that the decreased 5-HIAA accumulation is not a universal phenomenon in the group of vital depressions, either. We found subnormal values, i.e., values below the range of variation in the control group, in some 40% of cases (Van Praag and Korf, 1971; Van Praag et al., 1973a). Patients with and without demonstrable disorders of central 5-HT metabolism were indistinguishable in terms of psychopathological symptoms. From these facts we concluded that the group of vital depressions, although it tends to be homogeneous in psychopathological terms, is heterogeneous in biochemical terms, and comprises patients with as well as without disturbances in central 5-HT metabolism.

Our findings were confirmed by Goodwin (1976) and, in a CSF study without probenecid, by Bertilson and Åsberg (1975). In a group of patients with endogenous depression, both groups reported a bimodal distribution of CSF 5-HIAA values.

We have no indication that the presence or absence of an apparent central 5-HT deficit is related to the anxiety level. We were particularly interested in this question because in dogs (pointers) of a nervous strain, the 5-HIAA accumulation in the CSF after probenecid loading was found to be lower than in normal congeners (Angel et al., 1976).

That disparate processes can produce the same syndrome is not an unknown phenomenon in medicine. On the whole, all jaundiced patients show the same symptoms, yet the pathogenesis of the syndrome can be widely different. Another example is the anemia syndrome. Differences in pathogenesis could also exist in the vital depressive syndrome. This would have important therapeutic implications.

Cause or effect? It is of course a cardinal question whether the disturbances observed are *primary* or *secondary*; whether they contribute to the development of a depression, or result from it. In human individuals, conclusive information on this question can be obtained in only two ways: by establishing whether the metabolic disturbances preceded manifestation of the depression or followed it, and by establishing whether abolition of the metabolic disorder causes total or partial subsidence of the depression. The former method is virtually impracticable with human beings. The second is feasible: for example, 5-HT synthesis can be increased with the aid of 5-HT precursors. That the 5-HT precursor 5-HTP does indeed produce a therapeutic effect in 5-HT-deficient patients (Van Praag et al., 1972, 1974) will be discussed in the following chapter. This is indicative of the primary character of the metabolic disorder.

Central DA Turnover in Depressions

Decrease of DA turnover. Like the 5-HIAA concentration, the HVA concentration in lumbar CSF increases in response to probenecid administration. The accumulation was averagely decreased in a group of patients with vital depressions who were compared with normal test subjects and patients with personal depressions, psychotic reactions and neurotic personality disorders, without marked depressiveness (Roos and Sjöström, 1969; Van Praag and Korf, 1971; Sjöström and Roos, 1972; Van Praag et al., 1973; Goodwin and Post, 1974). The decrease was more pronounced in vital depressions with a unipolar course than in the bipolar types (Goodwin et al., 1973). We ourselves established that the diminished HVA accumulation is not a characteristic of vital depression per se, but is related to the state of motor retardation (Van Praag and Korf, 1971; Van Praag et al., 1973a). Since retardation is much more common in vital than in personal depressions, the disturbance in DA metabolism *seems* to be a characteristic of vital depressions. However, the phenomenon also occurs in personal depressions in which motor retardation is a prominent feature.

Diminished HVA accumulation after probenecid indicates a decreased intracerebral DA turnover. An additional argument in favor of a disorder of DA turnover was supplied by Sjöström and Roos (1972). They administered methylperidol as well as probenecid to a number of test subjects. Methylperidol is a neuroleptic of the butyrophenone series, which causes intensive stimulation of the central DA turnover, probably secondary to DA receptor block. Given undisturbed DA synthesis, the probenecid-induced HVA accumulation in CSF can be expected to show a marked increase in response to this agent. This was in fact observed, except in the vital depressive group, in which the additional increase caused by methylperidol was much smaller.

Cause or effect? The question arises whether the decreased DA turnover in retarded depressions is a primary (causative) phenomenon or a secondary one, e.g., a result of reduced motor activity and therefore, insufficient mixing of the CSF. I consider the latter to be unlikely: when depressive patients are given l-DOPA (a compound centrally converted for the most part to DA), those with normal HVA accumulation show no behavioral effect, but the low responders show normalization of the retarded motor activity (Van Praag, 1974). There is no distinct influence on the mood. This provides yet another argument in favor of an interrelation between DA and motor activity.

Central 5-HT and DA Turnover in Mania

Manic patients, like those with vital depressions, have been found to show a decreased accumulation of 5-HIAA and HVA (Sjöström and Roos, 1972; Post and Goodwin, 1974). This finding is the more remarkable because the so-called "mixing effect" acts precisely in the opposite direction: increased concentrations of 5-HIAA and HVA in lumbar CSF due to better CSF mixing as a result of hyperactivity (Post et al., 1973). The similarity in metabolic findings indicates that the same pathogenetic mechanisms can be involved in depression and in mania.

Also, the low HVA accumulation in the manic syndrome seems to contradict our finding that the low HVA accumulation in depressions relates to motor retardation. The contradiction, however, is not a cast-iron one. DA has two sites of predilection in the brain: the nigro-striatal and the mesolimbic system. The former system has a function in the regulation of motor activity; the latter possibly plays a role in the regulation of impulses and affects. It is therefore not inconceivable that a defect in DA metabolism might manifest itself both in motor functions and in affective functions.

Selectivity of the Probenecid Findings

Diminished 5-HIAA and HVA accumulation has been demonstrated mainly within the group of the vital depressions, but not (or only sporadically) in patients with personal depressions, psychotic reactions or nondepressive personality disorders. Thus far, therefore, it seems that the phenomenon is fairly selective within the range of psychiatric disorders.

Disorders of 5-HT and DA metabolism do occur in diseases which entail severe anatomical degeneration of brain tissue: first of all, in *Parkinson's disease*, in which HVA accumulation is markedly decreased (Olsson and Roos, 1968). This is an *in vivo* indication of the central DA deficiency initially demonstrated in post-mortem studies. It is the hypokinesia that shows the closest correlation to the DA deficiency and also the best response to l-DOPA. The decrease in HVA accumulation in retarded depressions is of the same order of magnitude as that in Parkinson patients. In view of these data, we formulated the hypothesis that a decreased intracerebral DA turnover is not a characteristic of a given nosological entity (Parkinson's disease) or of a given syndrome (vital depression); rather it is a characteristic of a given functional condition, probably that of hypokinesia, regardless of its etiology or the features of the syndrome within which this phenomenon occurs (Van Praag and Korf, 1971).

The 5-HIAA accumulation is likewise decreased in Parkinson's disease (Olsson and Roos, 1968). and this corresponds with the decreased 5-HT concentration found in the brain in post-mortem studies (Bernheimer and Hornykiewicz, 1964). The significance of this metabolic disorder for motor pathology is obscure. We have been unable to establish a correlation with the mood level (Puite et al., 1973).

Decreased 5-HIAA and HVA accumulations are found also in *presenile dementia, type Alzheimer* (Gottfries and Roos, 1973); the findings corresponded with the patient's degree of dementia. It is therefore plausible that the decreased 5-HT and DA turnover is based on involvement of serotonergic and DA-ergic systems in the degenerative process. Post-mortem studies could verify this hypothesis, but have not so far been carried out. The same applies to *multiple sclerosis*, in which low 5-HIAA and HVA accumulation values have also been found. In these cases, however, another explanation is to be considered. The phenomenon occurred only in very severely disabled, largely bedridden MS patients (Sonninen et al., 1973), but not in the less severe cases. It might well be, therefore, that immobility and poor CSF mixing rather than decreased 5-HT and DA turnover were the cause of the low accumulation in these cases.

The question of the specificty of disorders of the 5-HT and DA metabolism has not yet been exhaustively studied. With regard to future research into this question, I should like to make one further remark. The starting point in studies aimed at the cerebral substrate of neurological and psychiatric abnormalities is often a given clinical entity (vital depression, Parkinson's disease), i.e., a group of disorders in heterogeneous functions brought under the same denominator because they are usually observed together in the clinical setting. The dysfunction or group of interrelated dysfunctions (hypokinesia, rigidity, mood abnormalities) is much less commonly taken as a starting point in these studies. Is is highly possible, however, that metabolic (as well as physiological) disturbances correlate much more closely with certain motor, sensory or mental dysfunctions than with nosological entities or syndromes which represent a whole range of such dysfunctions. The concept of symptomatological specificity of a given metabolic disorder has been neglected in biological psychiatry and experimental neurology.

Reorientation on the symptom as a well-defined dysfunction—be it motor, sensory or mental—could be a productive stimulus for these areas of research (Van Praag, 1972; Van Praag et al., 1975).

Are the Disorders of 5-HT and DA Metabolism Syndrome-Dependent or Syndrome-Independent?

The question as to whether the disorders of 5-HT and DA metabolism are primary features which contribute to the occurrence of the depression, or secondary features, which result from it, can in my view perhaps be answered in favor of the former possibility. This view is based on the fact that compounds which promote 5-HT and DA synthesis exert a beneficial influence on certain depressive symptoms. Another question is whether the occurrence of the biochemical symptoms is dependent or independent of the depressive syndrome. If they are syndrome-dependent, primarily or secondarily, then they disappear together with the depressive symptoms. In the opposite case there are two possible explanations: (a) the biochemical disorders are unrelated to the depression, and the therapeutic effect of the precursor is not based on abolition of the MA deficiency; (b) the biochemical defect is related to the depressive syndrome, but in a predisposing rather than in a causative sense; it renders the individual more susceptible to the occurrence of depressions, but as such is not sufficient to cause manifestation of the depressive syndrome.

Decreased serotonergic transmission as a factor predisposing to depression and mania has been postulated as a theoretical model by Kety (1971) and Prange et al. (1974). They believed that the depression would become manifest when the CA-ergic transmission also diminishes, and that an abnormally increased activity in the CA system, combined with subnormal serotonergic activity, would underlie the manic syndrome (Fig. 11).

We reinvestigated a group of depressive patients after a symptom-free period of at least six months, with or without lithium prophylaxis. Such lithium prophylaxis as was given, was discontinued two weeks before the investigation. None of these patients had received antidepressants during the preceding three months. Two important findings were obtained. The HVA accumulation had increased after recovery in all low responders, and was now within normal limits. The data on the 5-HIAA accumulation were less unequivocal. The majority of low responders had shown an increase after recovery, but not always to a normal average level. In some patients, the accumulation had remained rather low, and three patients had shown no increase at all. Should persistent subnormal 5-HIAA accumulation indeed be a factor predisposing to manifestation of depression, then chronic medication with 5-HT precursors might have a preventive effect. The literature comprises a report on one patient who experienced a depressive phase every year over a 17-year period. During treatment with tryptophan in large doses, the expected depressions did not occur. She was considered to be emotionally much more stable than previously (Hertz and Sulman, 1968). We studied the prophylactic value of 5-HTP in 5 patients suffering from recurrent vital de-

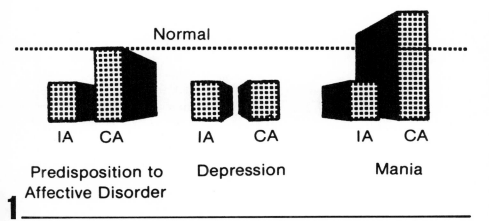

IA, Indoleaminergic Transmission
CA, Catecholaminergic Transmission

Possible relationship between central monoamine metabolism and the occurrence of depression and mania (from Prange et al., 1974).

pressions with (a) a high relapse rate and (b) a low 5-HIAA accumulation after probenecid during the depressive phase and after clinical recovery. The course of the depression was bipolar in three patients and unipolar in two patients. They were successfully treated with tricyclic antidepressants, and one month after discontinuation of this therapy were started on 200 mg 1-5-HTP and 100 mg MK 486 (a peripheral decarboxylase inhibitor) daily. After one month, the 5-HIAA response to probenecid was found to be amply above normal—a sign of their ability to convert 5-HTP to 5-HT. No relapse occurred over a period of 8-12 months. In view of the history of these patients, this was a striking result, which, however, will have to be verified by interposing placebo periods; this will be done. No similar study has so far been made in manic patients.

Conclusions

The probenecid-induced accumulation of 5-HIAA and HVA in the lumbar CSF is a better yardstick of the central turnover of 5-HT and DA than the base-line level of these metabolites, even though the probenecid technique is far from flawless. Several investigators have established that the 5-HIAA and HVA accumulation in response to probenecid can be decreased in depressions. This is indicative of a decreased turnover of the mother amines in the CNS. The 5-HT defect was most pronounced in vital depressions with a bipolar course, some investigators maintained that the DA defect was most pronounced in the unipolar group, but others believed that the DA defect correlates with motor retardation, regardless of the type of depressive syndrome in which this symptom occurs. According to Van Praag et al., the 5-HT defect is not a universal phenomenon within the category of vital depressions, either, but occurs in only 40-50% of cases. They postulated the existence of two pathogenetically different types of vital depression: a type with and a type without a demonstrable defect in 5-HT metabolism.

Since abolition of the suspected 5-HT and DA deficiency with the aid of precursors causes alleviation of the depressive syndrome or some of its features, the metabolic disorders probably play a role in the pathogenesis of the depressions. A fact which at first glance seems inconsistent with this is that, after disappearance of the psychopathological symptoms, the disorders of the 5-HT metabolism do not always, or not entirely, disappear. It is quite possible, however, that the biochemical defects are not so much direct causative as predisposing factors.

The few studies so far carried out in manic patients indicate similarities to depression in that decreased 5-HIAA and HVA accumulation values have been found.

Defects in central 5-HT and DA turnover have also been established in degenerative diseases of the CNS, e.g. , Parkinson's disease, Alzheimer's disease and severe cases of multiple sclerosis. They are possibly a result of involvement of MA-ergic systems in the degenerative process. We do not know to which psychological disorders the MA-ergic dysfunctions lead in these cases.

CHAPTER XI

Verification of the MA Hypothesis with the Aid of Drugs
MA Precursors and Synthesis Inhibitors

Justification of the Strategy

With the aid of Drugs, transmission in the central MA-ergic synapses can be influenced in two ways: *directly,* i.e., by influencing, say, the uptake or release of the transmitter or the susceptibility of the postsynaptic receptors, or *indirectly,* i.e., via interference with the synthesis or degradation of MA. These two strategies, so far as they are acceptable for use on human individuals, have been applied also to patients with affective disorders in order to test the MA hypothesis. The argumentation was as follows: If the MA hypothesis contains a kernel of truth, then drugs which increase the amount of MA available at the central postsynaptic receptors can be expected to produce an antidepressant effect, whereas drugs with the opposite action should promote depression. Conversely, it would reinforce the MA hypothesis if therapeutic methods with an unmistakable antidepressant effect, such as electroconvulsive therapy (ECT), could be shown to promote transmission, even though this is not a *conditio sine qua non* for the correctness of the hypothesis. It is conceivable, after all, that mood regulation can be influenced also via non-MA-ergic systems.

It is more difficult to give a prediction with regard to manic syndromes. If one regards depression and mania as counterparts, then an excess of CA or 5-HT should promote mania, while a decrease in these compounds should act against it. On the other hand, the few data available on the MA metabolism in mania are more indicative of pathogenetic similarities to depression than of extreme opposites.

This chapter discusses the "indirect strategy." The next chapter deals with the "direct strategy" and with the influence of certain therapeutic methods which influence mood (ECT, TRH, lithium) on central MA-ergic transmission.

Studies With Drugs that Increase 5-HT Synthesis

Tryptophan. The enzyme tryptophan hydroxylase is normally not saturated with substrate. Consequently, administration of tryptophan to test animals results in increased 5-HT synthesis, e.g., in the brain (Fernstrom and Wurtman, 1972). There are sound reasons for the assumption that conversion of tryptophan to 5-HTP takes place exclusively in serotonergic neurons (Lovenberg et al., 1968). This is a great advantage of tryptophan over 5-HTP, which can be decarboxylated in non-serotonergic neurons. In human individuals, the probenecid-induced accumulation of 5-HIAA in the lumbar CSF shows a marked increase in response to tryptophan (Dunner and Goodwin, 1972). This indicates that 5-HT synthesis is stimulated in the human organism also. Tryptophan therefore seemed a suitable agent to be used in efforts to establish whether an increase of cerebral 5-HT can have a therapeutic effect in *depressions*—even though the method is inefficient because only a small fraction of each tryptophan dose is utilized for central 5-HT synthesis.

The problem definition in the first tryptophan experiment was whether tryptophan can potentiate the therapeutic effect of a MAO inhibitor (in this case iproniazid) (Van Praag, 1962). A group of 20 vital depressive patients were treated for 10-14 days with 1,500 mg l-tryptophan, combined with a MAO inhibitor. The two drugs were simultaneously started in 10 patients, in the remaining 10, tryptophan was started after a few weeks' unsuccessful or insufficiently successful treatment with a MAO inhibitor. No potentiation of the specific effects of the antidepressant was observed. Coppen et al. (1963), who gave larger doses of tryptophan (214 mg racemic mixture per kg body weight), did observe an effect: depressive patients showed a better response to a combination of a MAO inhibitor (tranylcypromine) with tryptophan than to a combination of a MAO inhibitor with a placebo. Their findings were subsequently corroborated by several independent investigators (Glassman and Platman, 1969; Pare, 1963; Lopez-Ibor et al., 1973). Rats become hyperirritable and active in response to a combination of a MAO inhibitor with tryptophan—a phenomenon not observed after administration of tryptophan only (Grahame-Smith, 1971).

These data were not analyzed with a view to a possible influence of the type of MAO inhibitor used on the potentiating effect of tryptophan. It was recently established that MAO is not a simple enzyme but a complex of related enzymes which differ in substrate specificity. The effects of various MAO inhibitors on this system are not identical (Youdim et al., 1972). It is therefore conceivable that, in therapeutic terms, it makes a difference whether tryptophan is combined with one MAO inhibitor or with the other.

The findings obtained in depressions with tryptophan alone are certainly less perspicuous. Coppen et al., (1967, 1972) considered tryptophan to be a potent antidepressant, comparable in therapeutic potential with ECT and imipramine. Other investigators found no evidence of an antidepressant effect of l-tryptophan (Dunner and Goodwin, 1972; Bunney et al., 1971; Mendels et al., 1975; Carroll et al., 1970, Dunner and Fieve, 1975). All these studies were controlled studies. The findings reported by Herrington et al. (1974) lay between these extremes. They studied 43 patients with severe depressions, all candidates for ECT. Only one-half of this group did receive ECT, the other half being treated with 8 g l-tryptophan daily. Both groups had improved significantly after 2 weeks. Subsequently, however, the ECT group showed further improvement, but the tryptophan group did not. Assuming that the effects of both therapies were not based on nonspecific factors, and that no spontaneous improvement had occurred, these findings show that l-tryptophan is therapeutically active in severe depressions, but inferior to ECT. Finally, in a recent multinational study (Jensen et al., 1975), tryptophan was found as effective as imipramine in patients with endogenous depressions.

The tryptophan question can be described as still moot. I can discern three possible causes for this situation. To begin with, no attempt has so far been made to classify the depressive syndrome according to biochemical criteria. *A priori*, it seems improbable that the natural amino acid l-tryptophan could be a universal antidepressant. On the other hand, it cannot be excluded that a given category of depressive patients (those in whom a central 5-HT deficiency probably exists) might benefit from treatment with this agent. Mean values of a biochemically undifferentiated group of depressions would conceal such an effect (Van Praag and Korf, 1970, 1971, 1974). Secondly, there is uncertainty about the question whether certain tryptophan effects or their absence should in fact be ascribed to changes in central 5-HT metabolism. Finally, in each of the three experiments, tryptophan was combined with large daily doses of pyridoxine, which is a coenzyme in several tryptophan conversions. The doses given were many times as large as the therapeutic dose required in pyridoxine deficiency. In Parkinson patients, pyridoxine antagonizes the therapeutic effect of l-DOPA (Duvosin et al., 1969) by a mechanism which has not yet been entirely explained. It is therefore possible that pyridoxine has prevented an antidepressant action of tryptophan in these cases, but it is not likely: in rats, pyridoxine exerts no influence on the amount of tryptophan converted to 5-HT in the brain (Carroll and Dodge, 1971).

Even if it were established with certainty that tryptophan has no therapeutic effect on any type of depression, this would not automatically devaluate the hypothesis of a 5-HT deficiency in depressions. In fact, this deficiency might be based on a defect in synthesis, which makes the patient unable to utilize tryptophan for 5-HT synthesis.

Prange et al. (1974) studied the effect of l-tryptophan in *manic* patients: five received 6 g l-tryptophan daily, while the other five were given 400 mg chlorpromazine (Largactil). Both compounds produced a therapeutic effect, and l-tryptophan was the more effective of the two. This is a significant observation, which corresponds with the fact that indications of a reduced 5-HT turnover have been found in mania as well as in certain types of depression. This remarkable fact was confirmed by Murphy et al. (1974).

5-Hydroxytryptophan. Unlike tryptophan, 5-HTP is converted to 5-HT almost in its entirety. Consequently, it is a much more effective means of enlarging the amount of 5-HT in the brain.

Only a few studies have been made with 5-HTP, and their results were initially negative. Both alone and in combination with a MAO inhibitor, it failed to produce an antidepressant effect (Pare and Sandler, 1959, Kline et al., 1964; Glassman, 1969). However, the doses of 5-HTP administered were small, and CSF studies of 5-HT metabolites were omitted. Persson and Roos (1968) described one patient who showed a favorable response to 5-HTP. The striking feature of this case was that the patient had been resistant to ECT and amitriptyline medication.

Van Praag et al. (1972) gave larger doses of the racemic mixture of 5-HTP (3 g daily), i.e., the maximum tolerated by human individuals, and investigated two questions: (a) whether the agent was of use in the treatment of vital depressive patients, and (b) whether this was true in particular for patients with a subnormal 5-HIAA response to probenecid—a phenomenon considered to be indicative of a central 5-HT deficiency. The findings showed that 5-HTP probably does have a therapeutic effect in patients with a subnormal 5-HIAA response to probenecid (Van Praag et al., 1972), though the therapeutic results were not spectacular. In this respect, Van Praag's study differed from that described by Sano (1972), who in a large group (n = 107) of patients with unipolar and bipolar depressions observed striking results after relatively small doses of 1-5-HTP (50-300 mg daily), without addition of a peripheral decarboxylase inhibitor. The effect was quite pronounced and often ensued before the 4th day of medication. Van Praag et al. did, but Sano did not, use placebo controls, nor did Sano make CSF determinations. The data of Van Praag and Sano were confirmed by Takahashi et al. (1975): 5-HTP proved to be particularly efficacious in a subgroup of unipolar depressives.

In a small group of 6 patients with psychotic depressions, Brodie et al. (1973) were unsuccessful with 5-HTP. The patients were not classified on biochemical criteria, moreover, a negative selection was involved: none of them had responded to ECT or antidepressant medication. Trimble et al. (1975) reported that 1-5-

HTP given by intravenous drip had a positive effect on mood in nondepressive test subjects.

The above data certainly warrant further research into 5-HTP in depressions; it seems essential to me that further studies should involve probenecid tests, the results of which should be included in the analysis of results.

5-Hydroxytryptophan in combination with clomipramine. If the therapeutic effect of 5-HTP in depressions reported by Sano (1972) and Van Praag et al. (1972) is based on an increase of the amount of 5-HT available at the central postsynaptic 5-HT receptors, then it is to be expected that this therapeutic effect should be potentiated by drugs which produce a similar biochemical effect. One such drug is clomipramine (Anafranil). This is an inhibitor of the re-uptake of 5-HT and NA, just like other tricyclic antidepressants, but a relatively selective one: the re-uptake of 5-HT is strongly inhibited, but that of NA is only slightly so. The selectivity toward 5-HT also applies to man (Bertilsson et al., 1975). In test animals, clomipramine does potentiate the motor activation caused by 5-HTP in combination with a peripheral decarboxylase inhibitor (Modigh, 1973).* A similar effect has been observed in human individuals: the antidepressant effect of 5-HTP is potentiated by clomipramine (Van Praag et al., 1974). Clomipramine was given in daily doses of 50 mg, which as such are suboptimal. This was done to minimize this antidepressant's own effect. The study in question was open, and is now being verified in a controlled set-up. The antidepressant effect of clomipramine is likewise potentiated by tryptophan (Walinder et al., 1975).

Studies with Drugs that Increase CA Synthesis

Possibilities. Tyrosine hydroxylase is normally saturated with substrate, and administration of tyrosine consequently does not lead to an increase of CA synthesis (Levitt et al., 1965). But this effect can be produced with l-DOPA. This CA precursor is chiefly converted to DA in the brain, while only a small amount is converted to NA (Butcher and Engel, 1969), probably because the enzyme DA-β-hydroxylase is practically saturated with substrate. In principle, the only possibility of increasing the NA concentration selectively is afforded by DOPS, an amino acid which does not occur in the natural state and which is directly decarboxylated to NA, without intermediary DA synthesis (Blaschko et al., 1950). This compound has not yet been tested in depressive patients. It has been used in Parkinson patients, but with little success (Birkmayer and Hornykiewicz,

*Peripheral decarboxylase inhibitors are compounds which inhibit the conversion of 5-HTP to 5-HT (and of DOPA to DA), but do not penetrate into the brain. Peripherally, the amount of circulating precursor increases, and this results in an increased supply to the CNS. The useful effect of the precursor given is thus greatly enhanced.

1962). However, the affinity of DOPS for aromatic amino acid decarboxylase is low, and the question therefore remains whether the amounts tolerated by human individuals are in fact capable of causing a substantial increase in central NA concentration.

Small doses of l-DOPA. The initial experience with DOPA in depressive patients was not encouraging. Pare and Sandler (1959) were the first to treat depressions with l-DOPA, given intravenously as well as orally, in small doses (275 mg orally and 12.5-25 mg intravenously) and for only a few days. They observed no therapeutic effect, nor potentiation of the effect of MAO inhibitors. In the first double-blind study, larger doses of dl-DOPA were given (1,200 mg daily) and over longer periods (3-4 weeks). Again, no therapeutic effect was observed, either with or without MAO inhibitor (Klerman et al. 1963). More cheerful reports do exist, but these were based on open studies in small groups (Turner and Merlis, 1964, Matussek et al., 1966). During the same period (1962-1967), the results of DOPA therapy in Parkinson's disease (with comparable doses) were likewise disappointing.

Large doses of l-DOPA. As we know, Cotzias et al. (1967) discovered that dramatic improvement could certainly be achieved in Parkinson patients with DOPA given in large doses (in grams instead of milligrams). For this reason, l-DOPA was tried in large doses in depressive patients also. The number of pertinent studies has so far been small. Their results, though not unequivocal, do point in a certain direction. Of the 16 patients treated with l-DOPA by Goodwin et al. (1970), four showed unmistakable improvement: they improved in response to DOPA, and relapsed when given a placebo. Each of the four showed motor retardation. Five other retarded depressive patients did not improve in response to DOPA. The experiment was a double-blind study, each patient serving as his own control, this is to say that placebo and DOPA periods alternated in a manner unknown to either patient or evaluator. Matussek et al. (1970) compared the effects of l-DOPA (combined with a peripheral decarboxylase inhibitor) and a placebo in two groups of depressive patients, all with evidence of retardation. In the l-DOPA group, 12 out of 18 patients improved, versus 5 out of 13 patients in the placebo group. The difference in favor of l-DOPA was not statistically significant. It is to be noted that the l-DOPA dosage was fairly low (150 mg daily).

l-DOPA and HVA response to probenecid. A third experiment started from a different point of view. Ten depressive patients were selected on the basis of a biochemical criterion, i.e., the HVA response to probenecid (Van Praag, 1974). This was in excess of 100 ng/ml CSF in 5 patients, and below this level in the other 5 patients. In the latter group, motor retardation (as measured with a 10-point rating scale) was more pronounced than that in the former. The two groups did not significantly differ in mood level. All patients were treated for 2 weeks with 1-DOPA and a peripheral decarboxylase inhibitor. At the end of this period, the motor retardation had virtually disappeared in the patients with a low HVA response. In patients with a normal HVA response, the motor status had remained unchanged; this was to be expected, because prior to therapy it had hardly been abnormal. The mood scores were not significantly influenced by 1-DOPA. In both groups, finally, the HVA response to probenecid during medication was about twice the pre-therapeutic response. The low responders, too, were apparently able to convert DOPA to DA.

Weiss et al. (1974) telemetrically measured motor activity in a group of depressive patients, and also determined the HVA response to probenecid. They found no correlation between these values. I should like to make three notes with regard to these observations: (a) the report does not indicate whether patients with or without pronounced motor retardation were studied; (b) even if there were a relation between low HVA response and motor hypoactivity, this would not automatically imply a positive correlation between activity level in general and HVA response to probenecid; (c) the probenecid test gives no more than an impression of the central DA turnover. Delicate relations between DA metabolism and behavior parameters probably escape this technique.

All in all, I tend to regard the available data as indicating that DOPA can have a therapeutic effect on the motor pathology in depressions, and that DOPA-sensitive patients can be identified on the basis of a biochemical criterion. Yet DOPA is hardly an asset in clinical practice: its influence on the mood is insufficiently marked for that. Moreover, its activating effect seems scarcely more pronounced than that of tricyclic antidepressants with an activating component. Theoretically, however, these findings are certainly important, for the fact that DOPA has a useful effect in patients with a low HVA response indicates that disorders of the DA metabolism—which the low HVA response indicates—are primary factors and really play a role in the pathogenesis of elements of the depression.

Imperfections of the Precursor Strategy

Tryptophan. Administration of tryptophan is an inefficient method of increasing the central 5-HT concentration. No more than 3% of the tryptophan in the diet is converted to 5-HT, and only a small fraction of this becomes involved in central 5-HT production. Most of this tryptophan is subjected in the liver to the influence of the enzyme tryptophan pyrrolase, which splits the indole ring; this is accompanied by formation of kynurenine, which, via a series of intermediary products, is converted to the B vitamin nicotinic acid. Some tryptophan is used for protein synthesis, and a small fraction is finally converted to tryptamine. It is to be borne in mind, moreover, that the pyrrolase activity increases in response to large amounts of tryptophan (Knox, 1951), and that in depressive patients this activity is often increased anyway as a result of an increased

concentration of circulating corticosteroids (Knox, 1955). In rats, administration of tryptophan with a pyrrolase inhibitor (allopurinol) does indeed
lead to a more marked increase in intracerebral tryptophan than is produced by
tryptophan only. On the other hand, it is nevertheless doubtful (Fernando et
al., 1975) whether this factor of increased pyrrolase activity is of any practical
significance. Coppen's data suggest that it is not. He found that allopurinol did
not potentiate the effect of tryptophan in depressions (Carroll, 1971). Be this
as it may, the useful effect of a tryptophan dose is certainly small.

The effect of tryptophan on the central MA metabolism, moreover, is not
"punctiform." Administration of tryptophan to animals leads not only to an increased 5-HT level but also to a decrease in CA concentration (Green et al.,
1962, Sourkes et al., 1961), possibly because tryptophan and tyrosine compete
for the same transport mechanism to convey them into the CNS. There are
indications that the same phenomenon occurs in the human organism (Van
Praag et al., 1973).

In view of the above, it is hardly surprising that there are several known
tryptophan effects which almost certainly are not based on 5-HT synthesis,
namely, sedation, prolongation of sleep and reduction of the period of REM
sleep. These effects are not antagonized by pCPA, an inhibitor of tryptophan hydroxylase (Wyatt et al., 1970; Modigh, 1973). Furthermore, 5-HTP (combined
with a peripheral decarboxylase inhibitor) induces activation rather than sedation, and an increase instead of a decrease in REM sleep time (Modigh, 1972;
Van Praag et al., 1972, 1974, Carroll, 1971).

With regard to the toxicity of unphysiologically large doses of tryptophan, I
refer the reader to a review published by Harper et al. (1970). The possible risks
should be given serious consideration before one decides to give tryptophan for
longer periods, e.g., prophylactically (see chapter X). In this context, Carroll
(1971) made special mention of the risk of carcinoma of the bladder. This type of
carcinoma can be induced in test animals with the aid of the tryptophan metabolites 3-hydroxykynurenine, 3-hydroxyanthranilic acid and 2-amino-3-hydroxy-
acetophenone (Boyland, 1958), and their excretion is increased after tryptophan administration.

Large doses of l-tryptophan can produce side effects such as nausea, a light-
headed sensation and blurred vision, but these symptoms seldom take a serious
form. After combination with a MAO inhibitor, side effects can be more incon-
venient: tremor, hyperreflexia, flushing, hyperhidrosis, paraesthesias, ortho-
static hypotension and nystagmus. A large single dose of l-tryptophan (5 g or
more) given to patients being treated with a MAO inhibitor can provoke disintegration reminiscent of LSD psychosis—a syndrome I have described in detail
in my thesis (Van Praag, 1962).

5-Hydroxytryptophan. A handicap of 5-HTP medication is that 5-HTP decarboxylase, the enzyme which converts 5-HTP to 5-HT, is not specific and also

regulates, for example, the conversion of l-DOPA to DA (Yuwiler et al., 1959). 5-HTP which enters CA-ergic neurons is thus converted to 5-HT; this supersedes CA from the stores and might function as a false transmitter.

Two arguments can be presented to justify the 5-HTP strategy. To begin with, the conversion of 5-HTP in non-serotonergic neurons is small when moderate doses are given. This has been established by histochemical and biochemical methods (Fuxe et al., 1971; Korf et al., 1974). The biochemical test arrangement was as follows. Rats whose raphe nuclei had been destroyed under stereotatic control, and intact rats, were both treated with small doses of l-5-HTP (12 mg/kg) combined with a peripheral decarboxylase inhibitor. Both groups showed increasing 5-HT and 5-HIAA concentrations, but the increase was much more marked in the intact rats than in those with a destroyed raphe system. This indicates that the conversion of 5-HTP to 5-HT takes place largely in serotonergic neurons of the raphe system. The second argument is that the human 5-HIAA response to probenecid greatly increases in response to 5-HTP, whereas the HVA response hardly increases (Van Praag and Korf, 1975d). This indicates an increased 5-HT turnover and an unchanged DA turnover.

Side effects of 5-HTP are largely gastrointestinal: nausea, vomiting and a heavy sensation in the upper abdomen. When we used 5-HTP without peripheral decarboxylase inhibitor, we found for the majority of patients an unmistakable threshold value at 3 g dl-5-HTP per day. Gastrointestinal side effects were quite regularly seen at higher dosages. Contrary to expectation, the gastrointestinal side effects were exacerbated rather than improved when we combined l-5-HTP (400 mg daily) with a peripheral decarboxylase inhibitor (MK 486, 150 mg daily). Since the side effects often occurred within 10-15 minutes of ingestion of the 5-HTP capsules and proved to be quite unrelated to the blood 5-HTP concentrations (determined by Dr. J. M. Gaillard, Psychiatric Clinic, Geneva University), a direct effect of 5-HTP on the stomach and proximal segment of the small intestine seemed likely. For this reason 5-HTP was then supplied in coated capsules, which pass the stomach unchanged and disintergrate only as they reach the small intestine (at pH = 8.6). The side effects were thus substantially reduced in frequency as well as in intensity.

l-DOPA. This CA precursor interacts not only with the central CA system but also with the 5-HT system. The DA concentration increases, whereas that of 5-HT diminishes (Everett et al., 1970). The last-mentioned phenomenon is based on several mechanisms (Carroll, 1971): (a) competition of 5-HTP and l-DOPA for the same uptake system in the brain (Bartholini and Pletscher, 1968); (b) competitive inhibition of the conversion of 5-HTP to 5-HT in that DOPA and 5-HTP compete for the same decarboxylating enzyme (Yuwiler et al., 1959); (c) penetration of l-DOPA into serotonergic neurons, where it is converted to DA, supersedes 5-HT from the stores and starts to function as a false transmitter (Ng et al., 1970). A similar process possibly takes place in NA-ergic neurons. The fact that the NA concentration nevertheless fails to decrease after administration of large doses of l-DOPA could possibly be explained by a compensatory increase of the NA synthesis—a phenomenon which has indeed been observed following large doses of l-DOPA (Gershon et al., 1970).

The human serotonergic system is probably also influenced by l-DOPA, for in response to this compound the CSF 5-HIAA response to probenecid diminishes, and this indicates a decreased 5-HT turnover (Goodwin et al., 1971).

What applies to tryptophan is also true for l-DOPA: the useful effect of a dose is small. Only a small fraction of the dose given is utilized for CA synthesis, peripherally as well as centrally. The compound is for the most part converted to 3-0-methyldopa (Calne et al., 1969). In mouse brains, less than 0.1% of ^{14}C-labeled dl-DOPA was recovered after 20 and 60 minutes in the form of ^{14}C-labeled CA (Wurtman et al., 1970a). Chronic l-DOPA administration, therefore, makes such severe demands on the available S-adenosylmethionine (SAM), being the principal methyl donor for 3-0-methyldopa (and HVA), that the daily uptake of methionine (about 10-15 mmole) becomes insufficient (Wurtman et al., 1970b), and choline (via betaine) starts to function as methyl donor. This happens at the expense of the acetylcholine production (Carroll, 1971). The foregoing prompted Wurtman et al. (1968) to suggest that inhibitors of O-methylation can perhaps enhance the useful effect of a DOPA dose.

The inference of the above is that it is unwarranted to ascribe all DOPA effects simply to increased availability of CA in the brain. The diversity of biochemical influences involved possibly also explains why the psychiatric side effects of large doses of l-DOPA are so widely varied (Goodwin, 1972). There are reports on psychotic reactions: delirious features and paranoid states with an intact level of consciousness. Depressions are fairly common, but (hypo-)manic syndromes are also seen. Of great interest is a report by Murphy et al. (1971) that l-DOPA provoked (hypo-)manic behavior in 9 of the 10 patients with bipolar depressions studied. However, these authors noted that the depression scores failed to improve during such a period. What was involved, therefore, was motor disinhibition rather than true mania (of which syndrome, in my view, the mood component constitues an essential feature). Finally, l-DOPA can provoke irritability and aggressiveness, but without significant change in the degree of anxiety. For the somatic side effects of l-DOPA, one may refer to literature on Parkinson's disease.

Combination of l-DOPA and MAO inhibitors can give rise to severe hypertension and cardiac arrhythmias (Schildkraut et al., 1963; Hunter et al., 1970). Such combinations are therefore to be avoided.

Studies with Drugs that Inhibit MA Synthesis

Inhibition of tyrosine hydroxylase. α-MT inhibits the enzyme tyrosine hydroxylase and thereby the first step in CA synthesis (Spector et al., 1965).

In principle, therefore, it is a key substance in the verification of the MA hypothesis, always assuming that in the human organism also the CA synthesis is blocked to a substantial degree. The latter is certainly true. As judged by the renal excretion of CA metabolites, an overall block of some 60% can be achieved with large doses: up to 2,500 mg daily (Sjoerdsma et al., 1965). The decrease in the HVA concentration in the CSF warrants the conclusion that inhibition of synthesis occurs in the brain as well (Brodie et al., 1971).

The first question which α-MT prompts is, of course, whether this substance produces depression or at least elements of the depressive syndrome. In hypertensive patients, in whom α-MT was tested on a limited scale, sedation was reported as a side effect (Sjoerdsma et al., 1965). Schildkraut (1975), who observed some of the patients thus treated, reported that they developed unmistakable signs of mild vital depression, with transient hypomanic disinhibition after discontinuation of α-MT. In subhuman primates *(Macaca speciosa)*, α-MT provoked changes in social behavior (Redmond et al., 1971): the animals became reticent, less active and less inclined to seek contact with congeners. Their facial expression was described as indicating lack of interest in the environment. These behavioral changes were considered comparable to human depressive states.

Contrary to these findings, schizophrenic patients given 2,000 mg α-MT daily for several weeks were found to be sedated but not depressed (Gershon et al., 1967; Charalampous and Brown, 1967).

As pointed out above, what the effect of α-MT in mania will be is difficult to predict. Brodie et al. (1971) found that α-MT reduced the manic symptomatology in the majority of cases, although its effect was inferior to that of lithium carbonate. In the pathogenesis of certain manic symptoms, therefore, hyperactivity of CA-ergic systems probably does play a role. An observation pointing in the same direction was that the stimulating and euphorizing effect of intravenously given dl-amphetamine, a substance which stimulates intracerebral CA-ergic activity, was arrested by α-MT (Jonsson et al., 1971).

The α-MT data, although not quite unequivocal, therefore do point in the direction of a relation between MA metabolism and mood regulation. The fact that the results have been unspectacular could be due to the dosages, which at best can have accounted for about 50% inhibition of CA synthesis. In addition, it is to be noted that α-MT has not been introduced as antihypertensive, in view of side effects. It is used exclusively in the preoperative treatment of the rare tumor known as pheochromocytoma. In fact, experience with this compound has been very limited.

Inhibition of DOPA (= 5-HTP) decarboxylase. The widely used hypotensive drug methyldopa (α-methyl-1-3, 4-dihydroxyphenylalanine) inhibits DOPA (and 5-HTP) decarboxylase and thus induces a decrease of cerebral CA and 5-HT concentrations (Sourkes, 1965). Since no CSF studies have been carried out, there are no indications whether this also applies to human individuals.

Several authors consider methyldopa to be a depression-provoking substance, very like reserpine. The incidence is estimated to be 4-6% (Johnson et al., 1966; Hamilton, 1968). The *British Medical Journal* (1966) described methyldopa as contraindicated in the treatment of patients with a history of depression. This statement was not based on a comparative study of several antihypertensive drugs. The possibility therefore remained that what was involved was not a pharmacogenic depression but a depression provoked by the awareness of suffering from a relatively serious illness.

Three subsequent studies failed to establish a relation between methyldopa and depression. Each of the three, however, invites some criticism. Pritchard et al. (1968) continued the medication for three months—a period which may have been too short for depressive symptoms to develop. The remaining two studies (Bullpitt and Dollery, 1973; Snaith and McCoubrie, 1974) concerned patients who for the most part had already had methyldopa treatment for years. Selection may therefore have been involved: patients who developed a depression in the initial phase can have discontinued the medication or reduced the dose to an acceptable level. Obviously, therefore, there is a need for prospective studies.

Inhibition of DA-β-hydroxylase. α-MT is therapeutically active in manic patients (Brodie et al., 1971). Being an inhibitor of tyrosine hydroxylase, it reduces the production of both DA and NA. The relative contributions of the two effects to the antimanic effect cannot be established. Consequently, there is a need for a substance which selectively inhibits the enzyme DA-β-hydroxylase and thereby the conversion of DA to NA. Such a substance has recently been introduced in the form of fusaric acid (Nagatsu et al., 1970), which reduces the central NA concentration while that of DA or its metabolites increases (Voigtlander and Moore, 1970). Since the concentration of MHPG in human CSF decreases in response to fusaric acid (Sack and Goodwin, 1974), a similar effect probably occurs in the human organism. *In vitro,* fusaric acid exerts no influence on the activity of tyrosine hydroxylase, MAO and aldehyde dehydrogenase; in this respect, it differs from disulfiram (Antabuse)—also a DA-β-hydroxylase inhibitor but one of low selectivity which, for example, does inhibit the last-mentioned enzyme. Nor does fusaric acid influence the release or uptake of CA (Nagatsu et al., 1970).

In the CNS, DA-β-hydroxylase probably occurs only in NA-ergic neurons (Axelrod, 1972). Its inhibition, therefore, will selectively block NA synthesis in this system and have no direct effect on DA-ergic neurons. Since the conversion of DA to NA is inhibited, DA accumulates in the NA-ergic system and probably starts to function as false transmitter (Thoenen et al., 1965). If this theory is correct, then two factors contribute to the diminution of NA-ergic activity (Sack and Goodwin, 1974): (a) decreased NA synthesis, and (b) increased production of DA in NA-ergic neurons, which starts to function as false transmitter.

When used in manic patients (Sack and Goodwin, 1974), fusaric acid was found to cause exacerbation in severe cases (patients not so much with mania as with psychosis with symptoms of disinhibition). There is no certainty that this effect is actually based on the medication. The condition may have been exacerbated simply because effective medication was withheld. In milder cases of mania fusaric acid had no or hardly any effect. These findings warrant two tentative conclusions: (a) The therapeutic effect of α-MT in mania is based on reduction of the production of DA, not on that of NA. This is consistent with our hypothesis that DA-ergic hyperactivity plays a role in the pathogenesis of symptoms of motor disinhibition (Van Praag et al., 1975). (b) Elimination of the NA-ergic system induces psychotic symptoms. This corresponds with the hypothesis advanced by Stein and Wise (1971) that chronic schizophrenia is related to degeneration of NA-ergic neurons—a hypothesis supported by their finding that the activity of DA-β-hydroxylase in the brain is decreased in such patients (Wise and Stein, 1973). There is also an opposing hypothesis, which in fact postulates an excess of NA-ergic activity as important in the pathogenesis of psychoses (Snyder, 1972; Van Praag, 1975).

Inhibition of tryptophan hydroxylase. pCPA is a strong and relatively selective inhibitor of the enzyme tryptophan-5-hydroxylase, which therefore blocks the synthesis of 5-HT (Koe and Weissman, 1966). As judged by the renal excretion of 5-HIAA, 3-4 g pCPA daily reduces the overall 5-HT synthesis by 50-90% in normal test subjects (Cremata and Koe, 1966), and by 72-88% in carcinoid patients (Engelman et al., 1967). The CSF 5-HIAA concentration likewise decreases, indicating central 5-HT synthesis to be blocked also (Goodwin et al., 1973). The crucial question of the behavioral effects of pCPA in these cases cannot be unequivocally answered. Because of its toxic side effects, this compound has been therapeutically used only in patients suffering from a carcinoid: a malignant growth which usually produces large amounts of 5-hydroxyindoles. In the first systematic study of pCPA used for this indication, psychological effects "ranging from depression to hallucination" were observed, but it was added that it was difficult to assess these symptoms because of the poor physical condition of these patients (Engelman et al., 1967). Subsequent studies failed to add new points of view. This means that an unequivocal depressogenic effect of pCPA has not been established in carcinoid patients. In subhuman primates *(Maccaca speciosa)*, the CA synthesis inhibitor α-MT provoked a series of behavioral changes reminiscent of depression, as already mentioned; pCPA did not (Redmond et al., 1971).

The pCPA data so far collected are not consistent with the MA hypothesis (assuming that the pCPA dosage was sufficient for a 5-HT deficiency with functional implications); it is to be noted in this context that, owing to its toxicity, pCPA has never been given for longer periods to depressive or manic patients, nor to normal test subjects. The above conclusion is subject to yet another restriction: in a sophisticated study, Shopsin et al. (1974) demonstrated that pCPA arrested the therapeutic effect of imipramine within a few days, whereas several weeks of α-MT medication failed to influence this therapeutic effect. This indicates that the central 5-HT metabolism in any case plays a role in the antidepressant action of imipramine.

Conclusions

Precursor therapy is intended to abolish a suspected central MA deficiency. Although these precursors are far from selective in their effects, it is fairly certain that they stimulate the MA turnover in human individuals also. The results of precursor medication in depressions are not unequivocal. This applies to tryptophan as well as to 5-HTP and l-DOPA. On the other hand, indications that they are not therapeutically inert are too numerous to warrant rejection of this strategy as ineffective and interpretation of the results as evidence against the MA hypothesis. Most precursor studies, moreover, have been made in biochemically undifferentiated groups of depressive patients. This might explain the variability of the results. The few data now available do indicate the likelihood that MA precursors, although far from being universal antidepressants, can be effectively used in certain biochemical subtypes of depression.

Research with synthesis blockers has shown the likelihood that intracerebral DA-ergic hyperactivity plays a role in the development of manic symptoms, and that inhibition of CA synthesis promotes motor retardation and possibly also a decrease in mood level.

There are no convincing indications that inhibition of 5-HT synthesis has a depressogenic effect. On the other hand, the fact that inhibition of 5-HT synthesis arrests the therapeutic effect of tricyclic antidepressants supplies an unmistakable indication of a relation between 5-HT and mood regulation.

To summarize: in my view, the results of studies with precursors and synthesis blockers do not devaluate the MA hypothesis but in fact support it on certain points, albeit not in a conclusive manner.

CHAPTER XII

Verification of the MA Hypothesis with the Aid of Drugs and Other Methods of Antidepressant Therapy

Direct Influencing of the Transmission Process

Disorders in Central MA Metabolism and Therapeutic Efficacy of Tricyclic Antidepressants

Strategy. It is fairly generally accepted that the therapeutic effect of tricyclic antidepressants relates to their ability to inhibit the central (re-)uptake of 5-HT and NA into the neuron, thus enlarging the amount of transmitter available at the postsynaptic receptors. Some of these compounds have a certain selectivity in biochemical terms. Clomipramine (Anafranil) exerts a pronounced in influence on the 5-HT "pump" but has little effect on that of NA. The reverse is true for such compounds as nortriptyline and desipramine. Tricyclic antidepressants are believed to exert little influence on the (re-)uptake of DA.

If the disorders in central MA metabolism discussed do indeed play a role in the pathogenesis of depressions, then it is to be expected that not all tricyclic compounds can be equally effective, but that a depressive patient with a presumed central 5-HT deficiency would benefit most from a (relatively) selective inhibitor of 5-HT uptake, whereas a patient with a suspected NA deficiency would respond better to a compound which chiefly inhibits NA uptake. An as yet rather limited number of investigators have tried to test hypotheses of this type.

Tricyclic antidepressants and MHPG excretion. In two separate studies involving 12 and 16 patients, respectively, Maas et al. (1972, 1973a) found that the therapeutic efficacy of imipramine and desipramine is higher when the pre-therapeutic MHPG excretion in urine is low than when it is normal or increased. The difference in therapeutic response could not be correlated to a psychopathological criterion, specifically not to differences in motor pathology. Moreover, the low MHPG excretors responded to d-amphetamine by an improvement of mood, whereas the normal MHPG excretors did not (in fact, their mood often was adversely affected by amphetamine). The patients who improved in response to imipramine, desipramine and d-amphetamine, showed a moderate increase in renal MHPG excretion; this was not observed in patients who showed no improvement.

Beckman and Goodwin (1975) corroborated the findings reported by Maas et al. in a group of patients with unipolar depressions. Remarkably, they found no overlap: all responders had a lower MHPG excretion than the nonresponders. Contrary to the findings obtained by Maas et al., this study showed that the MHPG excretion *de*creased in *all* patients during imipramine medication.

It is possible to interpret these findings in such a way that they fit into the framework of the MA hypothesis. Imipramine, and particularly desipramine, block the re-uptake of NA more markedly than that of 5-HT; it can therefore be suspected that they enhance the central activity of NA more than that of 5-HT. Via several mechanisms, amphetamines enlarge the amount of CA available at the postsynaptic receptors. Assuming that (a) decreased renal MHPG excretion is indicative of a decreased NA degradation in the CNS, and (b) a causative relation exists between decreased NA turnover and depression, it can be expected that it is in particular the low MHPG excretors that benefit from medication with drugs which potentiate the central CA.

The weak spot in this argument is the first assumption. There can be no doubt that MHPG is produced peripherally as well, and that peripheral factors also influence its excretion. The fact that no correlation was established between the MHPG concentration in the CSF and that in urine (Shopsin et al., 1973) likewise casts doubt on this assumption. In the study of Prange et al. (1972), the MHPG excretion had no predictive value, even though it is to be noted that they determined the pre-therapeutic MHPG excretion levels fairly shortly (5 days) after discontinuation of the preceding medication, so that this could have influenced the results.

In a small group of patients, Schildkraut (1973) found indications of a relation between MHPG excretion and therapeutic effect of amitriptyline. Depressive patients with high or normal excretion values showed a better response than patients with a low excretion level. The findings of Beckmann and Goodwin (1975) also pointed in this direction: responders showed a higher MHPG excretion level than nonresponders. There was not even any overlap between the two groups. Since amitriptyline is biochemically a broad-spectrum antidepressant, in that it more or less equally influences 5-HT and NA uptake, it is difficult to interpret these results in the terms of the MA hypothesis.

Nortriptyline and 5-HIAA concentration in the CSF. Asberg et al. (1972) studied the possibility of a relation between baseline CSF 5-HIAA concentration and therapeutic response to nortriptyline. They found the therapeutic effect of this agent to be less pronounced in patients with a 5-HIAA concentration of less than 15 ng/ml CSF than in patients with higher 5-HIAA levels, even though the plasma nortriptyline level was well within the therapeutic range in both groups.

These findings, too, can be made to fit the MA hypothesis. Nortriptyline is a relatively selective compound: its inhibitory effect on the 5-HT "pump" is slight, whereas that on the NA "pump" is marked (Carlsson et al., 1969a,b). If a low CSF 5-HIAA level is indicative of a central 5-HT deficiency, then it is to be expected that a compound with a mostly NA-potentiating effect cannot be very effective in these patients. They would probably benefit more from a 5-HT potentiating antidepressant such as clomipramine. The correctness of this hypothesis is yet to be studied.

CSF 5-HIAA and clomipramine. In accordance with the findings of Asberg et al., mentioned above, are those of Van Praag (1976). In 30 patients with vital depressions, the relation was studied between post-probenecid 5-HIAA accumulation (interpreted as an indicator of central 5-HT turnover) prior to medication and the therapeutic effect of clomipramine, a mainly 5-HT-potentiating antidepressant. The course of the depression was unipolar in 16 and bipolar in 7 patients, the remaining patients being in their first depressive phase. The drug was administered during 3 weeks at a daily dosage of 225 mg. A negative correlation was established between the two variables in the sense that the therapeutic effect of medication increased by as much as pretherapeutic 5-HIAA accumulation had been lower.

The tentative conclusion was reached that, if the central 5-HT turnover is diminished in depression, then correction of this biochemical disturbance leads to alleviation of depressive symptoms.

After 3 weeks, 8 patients were considered not to have improved at all. Clomipramine was discontinued in these cases and replaced by a placebo for one week. The placebo period was followed by 3 weeks of daily administration of 150 mg nortriptyline, a mainly NA-potentiating antidepressant. Placebo and nortriptyline were given in capsules identical to those in which clomipramine had been administered. Neither raters nor patients were aware of the switch to placebo and then to nortriptyline. Of these 8 patients, 5 were considered to have markedly improved at the end of the third week of nortriptyline medication.

It is logical to expect these nortriptyline-susceptible patients to be NA-deficient. However, CSF studies have failed to corroborate this expectation. Renal MHPG excretion was not determined in these patients. The finding of a decreased excretion would elegantly clinch the hypothesis.

Conclusions. The relation between MA metabolism and treatment results with compounds which influence central 5-HT and NA more or less selectively lend support to two concepts: (1) that disorders of central MA metabolism can play a role in the pathogenesis of (vital) depressions, and (2) that the group of vital depressions is a heterogeneous one in biochemical terms. At least two types of vital depression seem to exist; 5-HT deficiency plays a role in the pathogenesis of one, while NA deficiency is important in the pathogenesis of the other. These types of depression are indistinguishable in psychopathological terms.

The above-described research strategy awaits compounds that approximate a 100% selectivity of their influence on the (re-)uptake of 5-HT and NA. Selective inhibitors of the 5-HT "pump" are being introduced (Fuller et al., 1974; Wong et al., 1974), but have yet to be clinically tested.

Tricyclic Antidepressants Combined with MAO Inhibitors

Because it is believed to give rise to serious cardiovascular complications, nearly all textbooks seriously caution against this combination, however logical it may seem (for a review, see Schukitt et al., 1971). On the other hand, there have been a few reports describing the (clinical) application of this combination without serious side effects, with good results in otherwise refractory depressions (Winston, 1971; Schukitt et al., 1971; Sethna, 1974). The combination so far most thoroughly tested is that of amitriptyline (25-100 mg daily) with isocarboxazid (Marplan, 10-20 mg daily) or phenelzine (Nardil, 45 mg daily) as the MAO inhibitor.

It would of course be unjustifiable simply to credit the MA hypothesis with this synergism. The MAO inhibitor could interfere with the microsomal liver enzymes which degrade tricyclic compounds, permitting the plasma level to rise higher. In view of the preliminary results reported by Snowdon and Braithwaite (1974), however, this is unlikely.

MAO Inhibitors Combined with Reserpine

MAO inhibitors block the intraneuronal degradation of MA, and possibly also the (re-)uptake of MA through the neuronal membrane. Both processes are assumed to lead to an increase in the amount of MA available at the postsynaptic receptors. Reserpine blocks the uptake of MA into the synaptic vesicles, causing them to fall victim to MAO, so that their concentration diminishes. If MAO is inhibited, however, than reserpine increases the MA concentration in the axoplasm, and consequently the MA "leakage" to the synaptic cleft is believed to increase markedly. This mechanism has been suggested as an explanation of the fact that reserpine does not sedate MAO-treated animals, but in fact stimulates them (Shore and Brodie, 1957).

If a deficiency of one or several MA plays a role in the pathogenesis of depressions, then it may be expected that the therapeutic effect of MAO inhibitors is potentiated by addition of reserpine. We tested this hypothesis in 32 patients with vital depressions; all were under 45 years old and had a normal baseline blood pressure, which was carefully kept under control (Van Praag, 1962). The daily reserpine dosage was 1 mg orally; the daily dosages of the MAO inhibitors tested (iproniazid and isocarboxazid) were 100-150 and 30-45 mg, respectively. In 15 patients, the two drugs were started simultaneously; the remaining patients were given reserpine after 4-6 weeks' unsuccessful or insufficiently successful medication with a MAO inhibitor. The combined therapy lasted 4 weeks. Eight patients of the second group, aged 25 to 40, were again given reserpine (5 mg, by slow continuous drip) 1-2 weeks after discontinuation of oral reserpine, while the MAO inhibitor was still being given; a careful check was kept on the blood pressure.

The result of this experiment can be summarized as follows. No overall therapeutic effect of reserpine was observed in the three groups as a whole. The combined treatment did not average a stronger or more rapid effect than the MAO inhibitor alone. In the refractory patients, addition of reserpine caused no *average* improvement. I emphasize the word "average" because a few patients (four in the first and four in the second group) showed a strikingly favorable response. They were indistinguishable from the other patients in psychopathological terms. They may have been recognizable on the basis of a biochemical criterion, but CSF studies were not yet being made at the time of this experiment. We have not repeated the experiment in view of the hypotensive risk.

Tricyclic Antidepressants and Central Stimulants

Amphetamine derivatives and methylphenidate (Ritaline) increase the amount of CA available at the central receptors via different mechanisms: they facilitate release into the synaptic cleft, and inhibit re-uptake and intraneuronal degradation (Carlsson et al., 1966; Rutledge, 1970; Sulser et al., 1968). In terms of the MA hypothesis, they can therefore be expected to potentiate the effect of tricyclic compounds. I know of only one study of this combination (Wharton et al., 1971), and this did disclose such a synergism. However, both methylphenidate (Perel et al., 1969) and amphetamines (Zeidenberg et al., 1971) inhibit the degradation of tricyclic antidepressants in the liver. This might also explain their potentiating effect.

Tricyclic Antidepressants Combined with Reserpine

Reserpine has also been combined with tricyclic antidepressants. Uncontrolled studies seemed to indicate success (Pöldinger, 1963; Hascovec and Rynasek, 1967), but controlled studies did not (Carney et al., 1969). In any case, this combination is hardly tenable theoretically. Tricyclic antidepressants are assumed to increase the extraneuronal MA concentration. Reserpine releases MA *within* the neuron, but these are degraded by MAO before they can "leak out" and produce an effect. There is consequently no theoretical ground for the hypothesis that reserpine could potentiate the therapeutic effect of the tricyclic compounds.

Enhancement of Central 5-HT Activity with the Aid of Chloramphetamines

4-Chloramphetamines and central 5-HT metabolism. In 1964, Pletscher et al. reported on a compound called 4-chloro-N-methylamphetamine (CMA), whose effect on the 5-HT metabolism tended toward selectivity. It causes a marked decrease in central 5-HT concentration without exerting any significant influence on the NA concentration. The 5-HIAA concentration likewise diminishes, giving rise to the suspicion that CMA inhibits 5-HT synthesis. However, tryptophan-5-hydroxylase in the liver was not inhibited by CMA. This suggested to Pletscher et al. (1966) that CMA releases 5-HT from the stores, but fails to cause an increase in 5-HIAA concentration because at the same time it is a (feeble) MAO inhibitor. No positive evidence of an alternative degradation of 5-HT was found.

It had meanwhile been established that tryptophan-5-hydroxylase in the brain was inhibited by CMA (Sanders-Bush et al., 1972) and that CMA reduces the 5-HT turnover (Sanders-Bush and Sulser, 1970; Korf and Van Praag, 1972). The two effects, probably interrelated, might be primary effects but also could be secondary, resulting from a feedback mechanism activated by excitation of postsynaptic 5-HT receptors. There were two phenomena indicative of greater likelihood of the latter mechanism because they suggested that CMA is more apt to increase than to decrease the activity in serotonergic systems.

To begin with, CMA was found to stimulate motor activity in test animals—a phenomenon observed also when central 5-HT activity is increased by means of 5-HTP in combination with a peripheral decarboxylase inhibitor (Modigh, 1972). Secondly, the motor-stimulating effect of CMA was inhibited by substances which reduce the amount of 5-HT available in the brain, e.g., cyproheptadine, a 5-HT antagonist and pCPA an inhibitor of 5-HT synthesis (Frey and Magnussen, 1968). Should CMA *primarily* block 5-HT synthesis, then cyproheptadine and pCPA could have been expected to potentiate the behavioral effects of CMA. The findings mentioned therefore suggest that the decrease in central 5-HT concentration in response to CMA is associated with increased serotonergic activity. A possible explanation of this paradox is that, in response to CMA, 5-HT is for the most part released, not *within* the neuron (where it would be inactivated before it could develop any activity), but in the synaptic cleft. In that case, the serotonergic activity would increase even though the absolute amount of 5-HT in the brain would diminish.

In the context of the MA hypothesis, a compound with a 5-HT-potentiating effect can be expected to exert an antidepressant influence. This is why between 1967 and 1970 we investigated the effects of CMA and a substance with a related action (4-chloramphetamine; 4-CA) (Fuller et al., 1965) in depressive patients (Van Praag, 1967). An exhaustive review of these studies is presented in two recent publications (Van Praag and Korf, 1973, 1975a).

It has meanwhile been established that the effect of chloramphetamines on the central 5-HT system is a very complex one, which comprises at least four components: (a) inhibition of 5-HT synthesis, (b) release of 5-HT from the synaptic vesicles, (c) inhibition of 5-HT transport through the neuronal membrane, and (d) MAO inhibition (Fuller and Molloy, 1974). However, these more recent findings have not upset the hypothesis that 4-CA enhances the central 5-HT activity in the acute experiment.

Chemical and psychopathological effects of 4-CA in human individuals. Of course, the foremost question was whether CMA and 4-CA also influence the human 5-HT metabolism and, more specifically, whether there are indications of a depletion of 5-HT stores. In response to both compounds, the renal excretion of 5-HT as well as 5-HIAA proved to show a gradual increase, and we interpreted this phenomenon as indicating that they do indeed interfere with 5-HT storage. The percentage of ^{14}C-labeled 5-HT converted to ^{14}C-labeled 5-HIAA did not diminish; overall MAO inhibition was therefore improbable. Finally, the overall 5-HT synthesis was assessed by successively loading test subjects with the 5-HT precursors (unlabeled) tryptophan and (^{14}C-labeled) 5-HTP and then measuring the fractions excreted in the urine as 5-HT and 5-HIAA. Tryptophan was not labeled because it is partly utilized in protein synthesis so that the radioisotope would long remain in the organism. Neither CMA nor 4-CA reduced the 5-HT and 5-HIAA yield, and consequently (overall) inhibition of 5-HT synthesis was unlikely. As judged by the response of CSF 5-HIAA to probenecid, CMA as well as 4-CA reduce the 5-HT turnover in the CNS.

All in all, therefore, it is quite likely that the chloramphetamines studied interfere with 5-HT storage without noticeably altering the overall ability to synthesize 5-HT. The decreased 5-HT turnover in the CNS could be secondary to 5-HT receptor stimulation. The conclusions from studies in human individuals are therefore in agreement with the findings of animal experiments.

When used in the treatment of depressive patients, CMA and 4-CA proved to be significantly superior in therapeutic effect to a placebo. They did not mutually differ in efficacy. Vital depression proved not to be an indication of preference, improvements were also seen in the group of personal depressions. This would seem to be a striking difference from the tricyclic compounds, with a definite indication of preference in the vital depressive group. Another striking finding was the absence of typical side effects as produced by nonchlorated amphetamine derivatives, e.g., agitation, insomnia, anorexia and hypertension. The side effects were found to be insignificant in general. In normal test subjects, too, the psychological effects of CMA were quite distinguishable from that of nonchlorated methylamphetamine.

We tend to interpret the results of this study as supporting the hypothesis that a central 5-HT deficiency plays a role in the pathogenesis of depressive symptoms.

Thyrotropin-Releasing Hormone (TRH)

In 1972, Prange et al. and Kastin et al. reported on the basis of independent studies that TRH very rapidly produces a therapeutic effect in depressions (within a few hours). The conventional antidepressants need 10-20 days to produce a therapeutic effect. In terms of chemical structure, TRH differs widely from the conventional agents with an antidepressant effect. It is a tripeptide produced in the hypothalamus, and already known to stimulate the hypophysis to release TSH (thyroid-stimulating hormone; thyrotropin), a hormone which enhances the hormone production in the thyroid. In animals, too, TRH induces behavioral effects. These probably result from a direct influence of TRH on the brain, for they occur in hypophysectomized animals also (Plotnikoff et al., 1972).

The results of reduplication studies were generally negative. Various investigators reported no therapeutic effect (Coppen et al., 1974, Montjoy et al., 1974; Benkert et al., 1974, Hollister et al., 1974) or only a marginal one (Van den Burg et al., 1975a,b; Takahashi et al., 1973). On the other hand, the various series studied included a few individual patients who showed a favorable if not readily defined response to TRH, but not to a placebo. Coincidence has not been established or ruled out with certainty. It would therefore be worthwhile to establish whether such patients can be psychopathologically or biochemically identified. Moreover, Wilson et al. (1973) observed an activating and euphorizing effect of TRH in normal women.

Assuming that TRH is not entirely devoid of an influence on mood and motor activity, the question arises whether (and if so, in what direction) TRH influences the MA metabolism. TRH has no effect on the intracerebral MA concentration (Reigle et al., 1974), nor does it change the intracerebral CA uptake (Horst and Spirt, 1974). However, it increases the CA turnover (Horst and Spirt, 1974, Keller et al., 1974, Constantinidis et al., 1974). Whether this is based on increased neuronal activity in CA-ergic systems, or represents a compensatory mechanism triggered by inhibition or block of transmission, remains to be established. Since TRH potentiates the motor-activating effect of l-DOPA combined with a MAO inhibitor in mice (Plotnikoff et al., 1972, 1974), the former mechanism is more plausible than the latter. In chronic experiments, however, the effect of TRH on the NA turnover was found to be no more than marginal (Reigle et al., 1974).

The effect of TRH on the 5-HT turnover is unknown, but it seems likely on indirect grounds that the 5-HT system is influenced. In rats treated with a MAO inhibitor, tryptophan induces hyperactivity—probably as a result of an abnormally high 5-HT concentration at the postsynaptic 5-HT receptors (Grahame-Smith, 1971). This effect is potentiated by TRH (Green and Grahame-Smith,

1974), and this is an indirect indication that TRH enhances central serotonergic activity.

Consequently, neither the clinical nor the biochemical books on TRH can as yet be closed. It would be regrettable if the largely negatively colored publications of the past few years were allowed to frustrate continued research.

Neuroleptics

Neuroleptics of the phenothiazine, thioxanthene, butyrophenone and butyl-phenylpiperidine type increase the intracerebral CA turnover (Andén et al., 1972, Nybäck and Sedvall, 1970). There are sound reasons to assume that this effect is secondary to postsynaptic CA receptor block (Van Rossum, 1967; Van Praag, 1975). However, the extent to which neuroleptics block DA and NA receptors, respectively, differs from compound to compound (Andén et al., 1970, Keller et al., 1973). The classification of neuroleptics on the basis of the ratio DA receptor-/NA receptor-blocking capacity does not coincide with their classification by chemical structure.

Neuroleptics of the groups of Rauwolfia alkaloids and oxypertine (Opertil), a neuroleptic with an indole ring, interfere with the uptake of MA into the synaptic vesicles, resulting in their increased degradation by MAO and a decrease in their concentration. Reserpine has no selective effect with regard to one particular MA, but oxypertine has: within a given dosage range, it has a predilection for the NA stores. The NA concentration decreases but the DA and 5-HT concentrations are less influenced (Hassler et al., 1970).

It can thus be stated in summary that all neuroleptics reduce transmission in one or several MA-ergic systems. In the context of the MA hypothesis, they can therefore be expected to have a depressogenic effect. Reserpine is indeed known to have such an effect (Goodwin, 1972); for the other neuroleptics, insufficiently investigated, this is dubious. The literature more or less regularly presents reports on depressions provoked by neuroleptics other than reserpine (Bohacek, 1973); but systematic studies with control groups are still lacking. Reports on depressions provoked by long-acting phenothiazine derivatives caused a measure of alarm (Alarcon and Carney, 1969) but could not be corroborated on the basis of controlled studies (Hirsch et al., 1973). Research into possible depressogenic effects of neuroleptics must beware of a number of pitfalls: (a) that depressive episodes can also occur in the natural course of a psychosis, (b) that extra-pyramidal side effects can be alarming and thus depressogenic, (c) that studies with neuroleptics invariably base themselves on their classification by chemical structure, although this gives no information on their biochemical action, as we have pointed out. It may well be that only some specific neuroleptics have a depressogenic effect, and that this is determined by their biochemical action profile. In that case, it would be unacceptable and deceptive to generalize about "the" neuroleptics and "the" phenothiazines (Van Praag and Korf, 1975b, Van Praag et al., 1975a).

An answer to this question might be supplied on the basis of controlled longitudinal studies of the mood effects of neuroleptics carefully classified also on the basis of their influence on the central MA metabolism.

Methylsergide

Methylsergide (l-methyl-d-lysergic acid butanolamide), a derivative of LSD, is a 5-HT (and tryptamine) antagonist (Doepfner and Cerletti, 1958). This has been established in the periphery. It antagonizes the vasoconstrictive effect of 5-HT as well as a series of 5-HT effects of extravascular smooth muscle tissue. The mechanism of this antagonism is reversible block of 5-HT receptors. We do not know whether methysergide also behaves as an antagonist of 5-HT in the brain. The compound has been used for years in the prophylaxis of migraine, but it is uncertain whether this effect is based on 5-HT antagonism.

Reports published by the end of the Sixties indicated that methysergide caused, within 48 hours, dramatic subsidence of manic symptoms (Dewhurst, 1968; Haskovec and Soucek, 1968; Haskovec, 1969). In some cases, these symptoms were in fact replaced by depressive symptoms (Van Scheyen, 1971). The same has been reported of another 5-HT antagonist: cinanserin (Kane, 1970). These were uncontrolled observations, and the findings are therefore questionable because the manic syndrome can markedly differ in intensity from day to day. The findings have not been corroborated in reduplication studies (Fieve et al., 1969; Grof and Foley, 1971; McNamee et al., 1972, McCabe et al., 1970). According to Coppen et al. (1969), methysergide can in fact cause exacerbation of manic symptoms.

The causes of these negative results can be widely varied. For example, manic symptoms may be unrelated to the functioning of central serotonergic systems; or the influence of methysergide on central 5-HT receptors may be too weak; or the doses given may have been too small (in any case, the gradient in methysergide concentration between blood and brain is great [Doepfner, 1962]); and so on. The pre-therapeutic CSF 5-HIAA concentration was not determined, nor was the possible influence of methysergide on this variable. Finally, it is possible also that the manic syndrome is heterogeneous in pathogenetic (biochemical) terms—much as it is now considered to be likely for the depressive syndrome—and that only a certain subgroup is responsive to methysergide. In that case, we must assume that this effect was concealed because group averages were used.

Lithium

Range of action. Lithium is one of the most striking new assets in pharmacopsychiatry: (a) it is therapeutically effective in manic patients, but much less in agitations of a different nature (Rimon and Räkköläinen, 1968; Lackroy and Van Praag, 1971); (b) it has a prophylactic effect in unipolar and bipolar depressions in that it reduces the number and/or severity of the phases (Schou, 1973); (c) it is alleged to be therapeutically effective in certain depressions. This last component of lithium action is the least certain. Its existence was initially doubted, but there are now indications that in particular the group of bipolar depressions could be lithium-susceptible (Goodwin et al., 1972). Moreover, there have been reports on two chemical parameters believed to be predictive in this respect: the capacity of the erythrocytes to take up lithium (increased in lithium-susceptible depressions [Mendels and Fraser, 1973]), and the increase or decrease of the plasma Mg and Ca levels during the first week of lithium medication, the lithium responders showing an increase (Carman et al., 1974). Our findings indicate the improbability of an antidepressant effect of lithium. Should lithium be therapeutically active in bipolar depressions, then recurrence during lithium prophylaxis could be expected to be more frequent in the unipolar than in the bipolar group. We have no indications that this is true.

Studies of the influence of lithium on the central MA metabolism have yet to lead to unequivocal results and conclusions. It is impossible to summarize these studies. Suffice it to mention a few findings in order to demonstrate that the influence of lithium on central MA is a very complex one, and that it may be very difficult to identify the net effect of this complex influence. For a more detailed review, I refer to publications by Schildkraut (1973) and Shaw (1975).

Noradrenaline. Lithium influences NA-ergic transmission in several different ways. There are some indications that it reduces the sensitivity of the postsynaptic NA receptor (Dousa and Hechter, 1970; Forn and Valdecasas, 1971). Moreover, in brain slices (Katz et al., 1968) and in isolated perfused cat spleen (Bindler et al., 1971), lithium reduces the release of NA provoked by direct electrical stimulation of the slices and stimulation of the splenic nerve, respectively. *In vitro,* finally, lithium increases the uptake of NA into synaptosomes (Baldessarini and York, 1970; Kuriyama and Speken, 1970). Each of the three effects mentioned is believed to result in a relative or absolute decrease of the amount of NA available at the corresponding receptors. Perhaps the increase in NA turnover observed after lithium administration both with the aid of tracer kinetics (Schildkraut et al., 1969) and by inhibition of degradation (Corrodi et al., 1967) may be regarded as an attempt to compensate the reduced receptor stimulation. However, the plausibility of this interpretation diminishes in view of the fact that, *in vivo,* lithium does not increase the NA uptake in the brain of test animals (Schildkraut et al., 1967, 1969) and that the influence on the NA turnover in chronic experiments soon diminishes (Corrodi et al., 1969), so that this effect cannot be used to explain the long-term effects of lithium.

Dopamine. Studies of the influence of lithium on the central DA system have been less exhaustive. The results, moreover, are contradictory. Some authors maintain that DA synthesis is decreased (Friedman and Gershon, 1973; Corrodi et al., 1969), whereas others hold that it remains uninfluenced (Ho et al., 1970). Of interest is the observation of Friedman and Gershon (1973) that the rate of DA synthesis decreases only after chronic lithium administration; this might correspond with the clinical experience that the therapeutic effect of lithium in manic patients does not become manifest until after several days to a week.

Serotonin. Lithium administered *in vivo* has no effect on the 5-HT uptake into brain synaptosomes (Kuriyama and Speken, 1970), but in brain slices it reduces the release of (radioactive) 5-HT following electrical stimulation (Katz et al., 1968). In view of these findings, one may be inclined to regard reduced availability of 5-HT at the central receptors as probable, just as for NA. However, this can hardly be reconciled with the data presented by Grahame-Smith and Green (1974): lithium combined with a MAO inhibitor made rats hyperactive, exactly as the combination of tryptophan and a MAO inhibitor does. The latter combination is believed to act via an increase of the 5-HT concentration at the postsynaptic receptors, and the authors regard it as likely that the same applies to the former combination.

The intracerebral 5-HIAA concentration increases in response to lithium (Schildkraut et al., 1969, Sheard and Aghajanian, 1970, Perez-Cruet et al., 1971), but it is uncertain whether this is based on an increased 5-HT turnover or on decreased transport of 5-HIAA from the brain. The available data on the influence of lithium on the 5-HT turnover are rather controversial. Mention has been made (but different methods of determination were used) of an increase (Sheard and Aghajanian, 1970, Perez-Cruet et al., 1971, Schubert, 1973; Poitou et al., 1974); of a decrease (Corrodi et al., 1969, Ho et al., 1970); and of absence of any change (Genefke, 1972).

Human studies. The findings obtained in human individuals are not consistent either. The renal excretion of MHPG increases in the initial phase of lithium medication, and then decreases (Schildkraut, 1973). The renal DA excretion likewise decreases (Messiha et al., 1970). The MHPG concentration in the CSF decreases (Wilk et al., 1972). The HVA concentration has been reported to increase (Wilk et al., 1972; Fyrö et al., 1975) or to decrease (Bowers et al., 1969). However, these studies encompass only a few patients, and probenecid responses were not studied. The same applies to the 5-HIAA concentration in the CSF, data on which are likewise confusing: most authors reported an increase during lithium medication (Mendels, 1971, Wilk et al., 1972; Fyrö et al., 1975), but Bowers et al. (1969) described it as unchanged.

An important study was published by Beckmann et al. (1975), in which they demonstrated biochemical differences between lithium responders and nonresponders. Comparing the predrug period with the third and fourth week of lithium treatment, all of the responders showed an increase in MHPG, while the nonresponders showed no change or a decrease. In addition, there was a tendency for the pretreatment MHPG excretion to be low in the patients who went on to show a clear-cut antidepressant response to lithium compared to those who were unequivocal nonresponders. Thus biochemical heterogeneity of the depressive disorders could explain the variability of the biochemical effects of lithium reported in the literature.

Some peripheral observations, too, are sufficiently interesting to be mentioned here. In human individuals, lithium reduces the hypertension provoked by NA infusions. The pressor response caused by tyramine is not affected (Fann et al., 1972). A possible explanation is diminished sensitivity of NA receptors. Another possiblity is that lithium promotes inactivation of NA, either through increased degradation or through increased neuronal uptake. An argument in favor of increased degradation can be found: lithium increases MAO activity in human blood platelets (Bockar et al., 1974). But increased re-uptake can also be argued: Murphy et al. (1969) demonstrated that lithium promotes the uptake of 5-HT into human blood platelets, and this cell type is considered to be a valid model of cerebral synaptosomes. These observations thus suggest that lithium reduces the amount of MA available at the receptors.

All in all, it seems justifiable to conclude that the influence of lithium on the MA metabolism is highly complex, and as yet insufficiently analyzed to be used in testing the MA hypothesis.

Electroconvulsive Therapy

Range of action. Electroconvulsive therapy (ECT) is an effective therapeutic method in depressions, and particularly in vital depressions (Carney and Sheffield, 1974). Owing to its rather drastic character, however, it has been largely replaced by antidepressant medication. The vital depression which shows no or no adequate response to antidepressants, and melancholia (deep vital depression with delusions—a syndrome which often shows no or no adequate response to antidepressants), have remained indications for ECT.

In my view, ECT is only of limited value in testing the MA hypothesis. It should be understood that the rush of current causes massive discharges in large areas of the brain and in the peripheral autonomic nervous system. During the convulsion, moreover, several endocrine glands are stimulated to secretion, and tonic and clonic contractions occur throughout large parts of the smooth and striated muscle systems (the last-mentioned factor can be reduced by establishing a neuromuscular block). The chemical homoeostasis of the organism is profoundly disturbed, and only an incorrigible optimist could hope to select from the numerous changes precisely those which determine the therapeutic effect.

Animal experiments. Nevertheless, the question whether ECT influences the central MA metabolism seems relevant in this context. Immediately after a single electroshock, the turnover rate of NA (Schildkraut et al., 1971; Schildkraut and

Draskoczy, 1974) as well as that of 5-HT (Engel et al., 1971; Ebert et al., 1973) in the rat brain is increased; however, the same phenomena are observed after all sorts of disagreeable interventions (Bliss et al., 1972; Thierry et al., 1968a,b). They can probably be regarded as nonspecific stress effects. Moreover, in a human individual a single shock is rarely effective: a series of shocks is required for a therapeutic effect. Therefore, the questions to be raised in this context are: Does the central MA metabolism change after a *series* of shocks? If so, how long does the change persist? The NA concentration is reported to remain unchanged under these conditions (Hinsley et al. 1968); that of 5-HT has been described as normal (Bertaccini, 1959) or as increased (Garrattini et al., 1960; Hinsley et al., 1968).

According to several authors, the rate of disappearance of intracisternally introduced ^{3}H-labeled NA increases after a series of shocks; moreover, a larger than normal fraction of it is converted to normetanephrine (Kety et al., 1967; Ladisich et al., 1969; Ebert et al., 1973), and tryosine hydroxylase activity increases slightly (Musacchio et al., 1969). This could mean that the NA turnover is increased and that more NA is released in the synaptic cleft (for NA methylation takes place extraneuronally). In addition, there are indications that the uptake of NA into synaptosomes is decreased after a single shock as well as after a series of shocks (Hendley and Welch, 1975). Extrapolated to an *in vivo* situation, this would mean an increased NA concentration in the synaptic cleft. These effects are demonstrable up to 3 days after completion of ECT, and they are more pronounced than those after other stress-producing interventions. Probably, therefore, they are shock-specific. Using the method of synthesis inhibition, however, Papeschi et al. (1974) found no influence of chronic ECT on the NA turnover.

The concentrations of 5-HT, 5-HIAA and the mother substance tryptophan are likewise increased for a few days after a series of electroshocks (Ebert et al., 1973). This might indicate an increased 5-HT turnover. However, the effect is not more marked than that after other stressors, and can therefore not be described as specific.

Thus the effects of repeated electroshocks in test animals are neither at variance with the MA hypothesis nor strongly in support of it.

Human studies. Few human studies have so far been carried out. The serum DA-β-hydroxylase activity increases during ECT (Lamprecht et al., 1974). Since this enzyme is released into the synaptic cleft together with the transmitter, this phenomenon might indicate that synaptic activity is in any case increased in the periphery. Stressors of a different nature, however, provoke the same phenomenon (Wooten and Cardon, 1973, Pflanz and Palm, 1973), and its specificity is therefore questionable. The CSF 5-HIAA and HVA concentrations do not change in response to ECT (Nordin et al., 1971). We ourselves found that the accumulation of these metabolites in the CSF after probenecid is increased on the day of ECT, regardless of its therapeutic effect. In patients whose 5-HIAA and HVA responses were diminished prior to treatment, normal responses were found one week after a successful ECT course. We have no data on patients who were refractory to ECT. In patients with normal pre-therapeutic responses, these values were found slightly but not significantly increased one week after ECT. In view of these data, it is likely that the 5-HT and DA turnover can increase in response to ECT, particularly in patients in whom the turnover of these amines was decreased to begin with. We have no data on the further course of 5-HIAA and HVA responses in these patients.

Conclusions

Transmission in MA-ergic synapses can be directly influenced with the aid of drugs, either in a facilitating or in an inhibitory sense. It has been shown that several procedures assumed to facilitate MA-ergic transmission probably relieve depression, although the results are not always unequivocal. This is in accordance with expectation. There is much more uncertainty about the psychological consequences of inhibition of transmission. For example, neuroleptics (assumed to be inhibitors of CA-ergic transmission) ought to be depressogenic according to the MA hypothesis. In view of clinical experience, this seems doubtful, yet almost no systematic studies have been made. The only 5-HT antagonist investigated in this respect had no demonstrable consistent effect on mood regulation.

Therapeutic methods with an established effect on mood regulation, e.g., ECT and lithium medication, could on the basis of the MA hypothesis be expected to exert an influence on the MA-ergic system. Such an influence is in fact demonstrable, but it is so complex and in part controversial that no net effect can be deduced from it and that, for the time being, it is impossible to use it in testing the MA hypothesis.

Direct influencing of MA-ergic transmission remains an important strategy in testing the MA hypothesis, but there is an urgent need for compounds which are more selective than the available substances. In this field, too, it seems to me to be of essential importance that in depression research classifications are made not only according to syndrome, etiology and course, but also according to pathogenesis, i.e., on the basis of biochemical points of view. The favorable effect of liver extract in pernicious anemia would probably not have been discovered if it had been given to all anemic patients, without distinction, and if the therapeutic results had simply been averaged.

CHAPTER XIII

Theories to Explain the MA Deficiency in Depressions

Possible Causes of the MA Deficiency

The MA hypothesis—the pros and cons of which have been discussed in previous chapters—postulates a functional cerebral MA deficiency as a (contributing) factor in the pathogenesis of certain types of depression. What can be the cause of this postulated deficiency? Generally speaking, there are three possibilities: precursor deficiency, decreased synthetic capacity, and decreased firing rate in the MA-ergic system. The firing rate in the presynaptic element is a factor which is among the determinants of the rate of MA synthesis. A decrease in firing rate is associated with a decrease in the activity of tryptophan hydroxylase, tyrosine hydroxylase and DA-β-hydroxylase. Less enzyme is produced, or its structure changes. In either case, the functional capacity diminishes, and MA synthesis decreases as a result (Thoenen, 1972; Bunney et al., 1973, Sheard and Aghajanian, 1968; Shields and Eccleston, 1972).

With these factors as a starting point, a number of theories to explain the MA deficiency have been advanced.

Precursor Deficiency

The amount of tryptophan in the brain is an important factor in the regulation of 5-HT synthesis. Intracerebral tryptophan deficiency leads to a decrease in 5-HT synthesis (Fernström and Wurtman, 1971). Coppen et al. (1973) found a

123

decreased plasma free tryptophan concentration in depressive patients. Moreover, the CSF tryptophan concentration was decreased (Coppen et al., 1972). The former finding was confirmed by Rees et al. (1974), and the latter was refuted by Ashcroft et al. (1973). In our patients, too, we found no decreased CSF tryptophan concentrations. As pointed out above (chapter VI. moreover, there are indications that depressive patients convert an abnormally large amount of tryptophan via the kynurenin route. This is believed to result from activation of tryptophan pyrrolase in the liver by circulating hydrocortisone, the amount of which is often increased in depressive patients. In test animals, tryptophan pyrrolase is so activated by hydrocortisone that the plasma tryptophan concentration decreases and the intracerebral 5-HT concentration diminishes by about 30%.

In view of these findings, it has been hypothesized (Curzon, 1969, Lapin and Oxenkrug, 1969) that depressions (or certain types of depression) involve primarily an increased secretion of adrenocortical hormone (possibly in response to increased ACTH production). This causes activation of tryptophan pyrrolase in the liver: more tryptophan is directed to the kynurenin route, and the plasma tryptophan concentration decreases. Plasma tryptophan (Tagliamonte et al., 1971), or rather the ratio plasma tryptophan/other neutral amino acids using the same transport mechanism to reach the brain (e.g., tyrosine, phenylalanine) (Fernström and Wurtman, 1972) is an important determinant of the intracerebral tryptophan concentration. It is therefore plausible that the intracerebral tryptophan concentration decreases and, with it, the 5-HT synthesis.

The theory seems to be fairly tenable at first glance. More careful consideration, however, reveals its weak spots: (a) as pointed out, administration of a single dose of hydrocortisone leads to a decrease in central 5-HT concentration, but this concentration is normalized as hydrocortisone administration is continued (Curzon and Green, 1968); (b) several authors have been unable to confirm the correlation between plasma tryptophan and intracerebral tryptophan concentrations (Morgan et al. 1975); (c) a high concentration of adrenocortical hormones in the plasma is not so much specific of depression as related to the factor anxiety (Sachar, 1967); (d) increased pyrrolase activity can be observed in a variety of conditions not associated with depression (Altman and Greengard, 1966), and nothing is known about pyrrolase activity in the liver in depressions, (e) there are investigators who have not measured increased kynurenin excretion in depressions (Birkmayer and Linauer, 1970). Consequently, it would be premature to regard the above-mentioned data as an outline of a biochemical *depression* theory. Perhaps they may later be found to have been more important for our understanding of the somatic substrate of anxiety.

Unlike tryptophan hydroxylase, tyrosine hydroxylase is normally saturated with substrate (Udenfriend, 1966), and it is therefore impossible to increase CA synthesis by supplying extra tyrosine. Yet it is quite conceivable that a tyrosine deficiency reduces CA synthesis, although there are hardly any indications that it really does. The peripheral data—plasma tyrosine concentration, baseline and after oral tyrosine loading—are inconsistent (Takahashi et al., 1968; Birkmayer and Linauer, 1970, Benkert et al., 1971). Moreover, the CSF tyrosine concentration is normal in depressive patients (Van Praag et al., 1973). There is no theory which relates the supposed central CA deficiency to tyrosine deficiency.

Diminished Synthetic Capacity

There are indications that the "tryptophan tolerance" is decreased in depressive patients. After oral loading with l-tryptophan, the blood tryptophan level was found to be above normal during the first few hours (Coppen et al., 1974). This may be an indication that the chemical "processing" of tryptophan, in whatever direction, is delayed; of course, other explanations are conceivable also. A comparable phenomenon has been observed in the CNS. During the first few hours following administration of a large oral dose of tryptophan, the CSF tryptophan concentration showed a more marked, and the 5-HIAA concentration a less marked increase than that in the control group (Van Praag et al., 1973b). This suggests that the tryptophan administered "congests" so to speak, in the CNS in depressive patients, and is less readily converted to 5-HT.

Since the conversion of 5-HTP to 5-HT is undisturbed (Coppen, 1967; Van Praag and Korf 1975d), these findings are indicative of a decreased tryptophan hydroxylase activity, or a diminished accessibility of the neuron to tryptophan.

There are no analogous data on the CA metabolism. In depressive patients, the ability to convert l-DOPA to HVA is undisturbed (Goodwin et al., 1970; Van Praag, 1974).

Postsynaptic Defects

In the accepted view, tricyclic antidepressants reduce the central re-uptake of 5-HT and NA into the nerve endings of serotonergic and NA-ergic neurons. The postsynaptic receptors are activated as a result. This leads to a reduced firing rate in the presynaptic element, and subsequently to a gradual decrease of tryptophan hydroxylase and tyrosine hydroxylase activity. The last-mentioned phenomenon is regarded as a side effect of the receptor stimulation. It is the receptor stimulation that is held responsible for the therapeutic effect.

A fact which it is difficult to reconcile with this view is that MA uptake is inhibited within a few minutes by tricyclic antidepressants (Schildkraut and Kety, 1967), whereas the therapeutic effect as a rule does not become manifest until after a latent period of 10-20 days. The theory formulated by Mandell et al. (1975) offers an explanation. They confined their reasoning to the NA-ergic system, and postulated that the postsynaptic NA receptors are hyperactive rather than hypoactive in depression, either due to hypersensitivity or as a result of a chronic NA surplus, e.g., due to reduced extraneuronal NA degradation. By way of compensation the presynaptic firing rate decreases, and tyrosine hydroxylase activity gradually diminishes as a result. This last phenomenon is reflected in low concentrations of CA metabolites in the CSF. Tricyclic anti-depressants expose the postsynaptic receptors to acute activation, and this results in further reduction of tyrosine hydroxylase activity. The latter effect is considered not to be an epiphenomenon but the key to their therapeutic effect: it gives the hyperactive (hypersensitive) receptor extra protection.

This is an interesting theory, but one that is not readily reconciled with various other phenomena. For example, although little is known about the psychic action of the CA synthesis inhibitor α-MT in human subjects, there is no indication at all that it could have an antidepressant effect. On the other hand l-DOPA (a CA precursor which enhances CA synthesis in depressive patients) is of therapeutic value in the treatment of these patients. Moreover, according to Mandell's theory, tricyclic compounds should initially cause exacerbation of the depression (receptor stimulation), but there is no evidence that they do.

Ashcroft et al. (1972) also encompassed the postsynaptic receptor in their arguments. On the basis of the observation that there are depressions with and without decreased concentrations of MA metabolites in the CSF, they distinguished two types of depression. In the one type (the so-called low-output depression), they believed the supply of stimuli to MA-ergic systems to be abnormally low, with a consequent low transmitter output and decreased concentration of amine metabolites in the CSF. In the other type (the so-called low-sensitivity depression), the primary factor was believed to be reduced sensitivity of the postsynaptic receptors with, at least in the initial phase, normal transmitter output and normal CSF concentrations of metabolites.

The receptor theories are purely speculative for the time being, because we have no yardstick of human receptor activity. Determination of cyclic AMP in the CSF is not likely to open important perspectives in this context (Sebens and Korf, 1975).

Conclusions

Efforts to explain the MA deficiency postulated in depressions have so far remained futile. There are no convincing arguments in favor of a precursor deficiency. There are some indications of reduced synthetic capacity, but they are insufficiently strong to carry any theory. Finally, there is the view that the defect is postsynaptically localized, rather than presynaptically, and relates to a changed sensitivity of MA receptors. Interesting as they may be, these theories are still entirely speculative.

CHAPTER XIV

Preliminaries to Expansion of the MA Hypothesis

Acetylcholine

In the quest for biological factors of importance in the pathogenesis of affective disorders, the central MA metabolism has so far been the cynosure of interest. This is not based on a conviction that MA-ergic systems play a more or less exclusive role in mood regulation. The predilection is a technical one, and not one of principle. Concentrations of MA and their metabolites can be determined without too much difficulty. The MA metabolism, moreover, is of a linear character: there is a mother substance and there are end products, and consequently it was not too difficult to evolve turnover standards. Compare this with a substance such as acetylcholine, also a central transmitter. Its determination is difficult, and often still dependent on laborious biological methods. Moreover, choline is a metabolite as well as a precursor of acetylcholine, and consequently the choline concentration cannot give unqualified information of acetylcholine consumption. Nevertheless, recent reports have contained some data which may well break the "MA barrier," for they suggest that cholinergic mechanisms are involved in the regulation of mood and motor activity.

Janowski et al. (1974) studied the influence of physostigmine on mood, psychomotor activity and impulse level in patients with depressive, manic or schizoaffective syndromes. The last-mentioned syndromes are mixtures of depressive, manic and schizophrenic components. Physostigmine inhibits cholinesterase and thereby increases the intracerebral acetylcholine concentration. The action of this compound was compared with that of a placebo and that of neostigmine—likewise a cholinesterase inhibitor, but one which does not readily enter the brain and therefore largely confines its activity to the periphery. In an effort to reduce the peripheral cholinergic effects of physostigmine and neostigmine, all test subjects were pre-treated with methylscopolamine, an anticholinergic which does not act centrally.

In all test subjects, physostigmine induces unmistakable motor retardation. No hypnosedative effects were observed, and the psychotic symptoms showed no aggravation. The motor retardation was reminiscent of that observed in retarded vital depression. In patients with mood disturbances (manic as well as depressive) moreover, the basic mood was observed to shift in a depressive direction. There was a spontaneous return to the initial situation within 30-60 minutes of physostigmine administration, and almost immediately after intravenous administration of an anticholinergic such as atropine. Neostigmine caused no discernible changes in behavior. The ability of physostigmine to cause motor retardation and lower the mood level was confirmed in an independent experiment with two marijuana users in a "high" state (El-Yousef et al., 1973). In manic patients, these effects could be made subservient to treatment: physostigmine reduced hyperactivity and agitation. In some patients, the baseline mood even overshot in a depressive direction (Janowski et al., 1973).

On these grounds, Janowski et al. (1972) postulated that certain cholinegic and CA-ergic systems are reciprocally related and that both could be involved in dysregulations of affect. Predominance of cholinergic over CA-ergic activity—relative (due to insufficient CA-ergic activity) or absolute—was believed to lead to motor retardation, a lowering of the mood level, and ultimately to depression. Relative or absolute predominance of CA-ergic over cholinergic systems was believed to stimulate motor activity and enhance the mood level, to end ultimately in mania. Thus a certain affective state was conceived of as determined by a given cholinergic/CA-ergic balance. The authors offered some additional arguments in support of this theory (for the pertinent literature, I refer to their publication). These arguments can be summarized as follows.

(1) Reserpine depletes the central MA stores, but is also a central cholinergic. The so-called "reserpine depression"—a syndrome characterized by reduced motor activity and despondency—has been ascribed to a central MA deficiency. Nothing, however, hampers the hypothesis that a cholinergic component is also of significance in this respect.

(2) Tricyclic antidepressants enhance, via inhibition of the re-uptake, the amount of MA available at the central postsynaptic receptors. This action is believed to underlie their antidepressant effect. However, these compounds are

also central anticholinergics, and there can be no objection to the theory that this action also contributes to their therapeutic effect.

(3) Insecticides and nerve gases of the cholinesterase type, i.e., substances which increase the intracerebral acetylcholine concentration, can give rise to negative mood changes and reduction of motor activity. In addition, however, they produce so many other psychological and somatic toxic symptoms that in my view this argument is not very convincing.

(4) Methylphenidate (Ritalin) is an amphetamine-like substance that stimulates human motor activity and causes either dysphoric or euphoric mood changes. These effects are probably related to increased intracerebral CA-ergic activity. They are abolished by physostigmine. Inversely, physostigmine-induced inhibition is immediately arrested by intravenously given methylphenidate.

(5) A central DA deficiency exists in Parkinson patients. The disease symptoms, and particularly hypokinesia, show a favorable response to enhancement of DA-ergic activity (l-DOPA) but also to attenuation of cholinergic activity (anticholinergics). This suggests a reciprocal relation between cholinergic and DA-ergic systems, at least in regulation of motor activity. Depressions are more frequent in Parkinson patients than in comparable controls. It is not inconceivable that cholinergic predominance plays a role in this respect.

The above-mentioned data open wider perspectives for theories on the pathogenesis of depressive syndromes. They break what may be described as the "MA monopoly." On the other hand, we should not lose sight of the fact that all data so far indicative of a relation between acetylcholine and depression are of an indirect nature. There are no direct arguments. These could be obtained only by CSF studies. The CSF is the sole source of information of metabolic processes in the human CNS. Whether it may yield data with an indicative value with regard to the central acetylcholine metabolism is still a moot point. It is on this point that research should now be focused (Flentge and Klaver, 1975).

Phenylethylamine

2-Phenylethylamine (phenethylamine; PEA) is a substance closely related structurally to amphetamine, which is found in human urine (Jepson et al., 1960) and in the human brain (Mosnaim et al., 1973) in virtually the same concentration as CA and 5-HT. Its renal excretion is alleged to be independent of diet (Jepson et al., 1960). Moreover, since PEA passes the blood-CSF barrier without difficulty, Sabelli and Mosnaim (1974) considered the PEA concentration in urine to be indicative of that in the brain—a bold postulate in that virtually nothing is known about the peripheral PEA metabolism.

PEA is formed from phenylalanine in response to aromatic amino acid decarboxylase, and degraded by MAO to phenylacetaldehyde, which is then oxidized to phenylacetic acid or reduced to phenylethanol. While CA and 5-HT are the predilected substrate of type A MAO, PEA is that of type B (Neff and Yang, 1974). Like amphetamine, PEA behaves like a central stimulant but has never been tested as an antidepressant because it produces too many side effects (Sabelli and Mosnaim, 1974).

The urinary excretion of PEA was found decreased in depressive patients by two groups of investigators (Fischer et al., 1968; Mosnaim et al., 1973; Fischer et al., 1972; Sabelli and Mosnaim, 1974). In "endogenous depressions," the phenomenon was observed in 100% of cases. Tricyclic antidepressants as well as MAO inhibitors increase the intracerebral PEA concentration (Fischer et al., 1972; Sabelli and Mosnaim, 1974). Tricyclic antidepressants—long believed not to influence MAO—were recently found capable of preferential inhibition of type B MAO (Roth and Gillis, 1974; Edwards and Burns, 1974).

These are the grounds on which Sabelli and Mosnaim (1974) advanced the hypothesis that a central PEA deficiency should be an important factor in the pathogenesis of ("endogenous") depressions. This hypothesis is a rather bold one in view of the fact that the function of PEA in the brain is unknown, and that no studies have been devoted to the central PEA metabolism in (depressive) human individuals. Nevertheless, it can serve an important function if it prompts an interest in those central amines that have so far received too little attention in this context.

Conclusion

It is unlikely that MA-ergic systems monopolize mood regulation. Nevertheless, the building blocks for a more comprehensive theory are still lacking. The cholinergic system has been suggested as a possible candidate, but only indirect arguments have been presented. Direct studies of the human cerebral acetylcholine metabolism are likely to pose quite substantial problems. The candidacy of phenylethylamine is even less strong, simply because nothing is known as yet about the function of this compound in the brain. Obviously, however, this does not alter the necessity of giving non-MA-ergic transmitter systems more attention than they have hitherto received.

General Conclusions

(1) With regard to the hypothesis that a relation exists between behavioral disturbances in affective disorders on the one hand, and disorders of the central MA metabolism on the other, I believe that arguments in favor outweigh arguments against. The principal relevant data were obtained by CSF studies, post-mortem examinations of the brain, and studies of the psychological effects produced by drugs with a more or less focused influence on the central MA metabolism. The material available indicates the probability that depression can be associated with a decreased turnover of 5-HT and CA; the manic syndrome—much less intensively studied—can be associated with a decreased 5-HT turnover with hyperactivity rather than hypoactivity of the CA production. According to the most widely accepted theory, the decreased MA turnover reflects a decreased transmission in MA-ergic systems, and this phenomenon is directly linked to the occurrence of depressive symptoms. There is an alternate view which holds that the CA-ergic system is in fact hyperactive in depression. This view interprets the decreased CA turnover as a compensatory phenomenon, aimed at limitation of hyperactivity. This theory is interesting, but speculative. It is yet to be established whether the suspected decrease in MA turnover is based on a deficiency of the mother substance, an enzyme defect or a primary reduction of the firing rate in the MA-ergic system.

(2) Disorders of the MA metabolism have been found not in all types of depression, but in those which are symptomatologically characterized by the so-called "vital" ("endogenous") syndrome, which take a unipolar or bipolar course, and which involve relatively little disturbance of the premorbid personality structure.

(3) It is plausible that disorders of the MA metabolism play a role in the pathogenesis of depressions, instead of resulting from them. This can be deduced from the fact that control of these disorders can abolish or alleviate the depressive syndrome or some of its elements.

(4) The disorders of the 5-HT metabolism do not always, or always entirely, disappear after clinical recovery from the depression. A possible interpretation of this fact is that the metabolic disorders represent a predisposing rather than a direct causative factor: they increase the individual's susceptibility to depression, so to speak. This would imply that abolition of a chronic 5-HT deficiency, e.g., with the aide of precursors, should have a prophylactic effect. Our preliminary findings indicate that this is probably true.

(5) There is a marked inclination to correlate biological factors found in behavioral disorders, e.g., in depressions, with nosological entities (e.g., manic-depressive psychosis) or syndromes (e.g., vital depression). The efficacy of this approach is questionable, and it may well be advisable to combine this procedure with one in which the syndrome's components (the disorders of psychological function) are analyzed and then correlated with biological functional disorders. In our studies of the DA metabolism, this proved to be an effective procedure. An increase or decrease in DA turnover was found not to be correlatable with a given nosological or syndromal entity (e.g., Parkinson's disease, depression, psychosis) but rather with the patient's motor status, i.e., the degree of motor retardation or agitation. In my view, it would certainly be unjustified to describe the metabolic disorders as nonspecific on these grounds. They are specific on a symptomatological, but nonspecific on a nosological and syndromal level, simply because the symptoms in question happen to be features of several different syndromes. In this context, I define the term "symptom" as a well-defined disturbance of a psychological function.

(6) There is a great need for drugs with an optimally selective and circumscribed influence on given MA-ergic systems—not only for further verification of the MA hypothesis but also, and to an equal extent, because I suspect that in this way the treatment of depression could be refined and better adjusted to the individual symptom pattern.

(7) The data available would seem to suggest that depressions are not a homogeneous group in pathogenetic (biochemical) terms either; that in fact biochemical differences can exist even within a group which tends toward homogeneity in psychopathological terms. Within the group of vital depressions, for example, patients with and without demonstrable disorders of the central 5-HT metabolism can be found. In the study of depressive patients, particularly if aimed at testing methods of medication, data on the MA metabolism could advantageously be included in the analysis of results. Averages of results in biochemically undifferentiated groups can have the effect of a veil. This implies that analysis of MA metabolites in lumbar CSF will have to become a routine part of biologically oriented depression research (the CSF affords data of greater relevance to the functioning of the CNS than the periphery can yield). CSF studies after probenecid loading are far preferable to determination of baseline concentrations in CSF, because the probenecid test greatly enhances the instructive value of CSF studies.

(8) There is no reason to assume that the pathogenesis of depressions could be understood exclusively in terms of disturbed MA metabolism. It is therefore by all means advisable to include other transmitter systems in research. The central cholinergic system is an obvious candidate in this respect, even though it will be far from simple to obtain information of this system in living human individuals.

(9) Quite apart from the fact that it probably contains a kernel of truth, the heuristic value of the MA hypothesis for psychiatry cannot be overestimated. It has catalyzed an impressive amount of research, not only in the biochemical field but also in pharmacotherapy, in psychopathology and in experimental behavior research. The MA hypothesis demonstrates that multidimensional but target-oriented experimental research is quite feasible in psychiatry as long as a plausible working hypothesis is available. It is in these terms that the MA hypothesis possibly has its greatest importance.

References

Abdullah, Y.H., and Hamadah, K. (1970): 3'5' - Cyclic adenosine monophosphate in depression and mania. *Lancet, 1:* 378-381.

Alarcon, R. de, and Carney, M.W.P. (1969): Severe depressive mood changes following slow-release intramuscular fluphenazine injection. *British Medical Journal, 3:* 564-567.

Altman, K., and Greengard, O. (1966): Correlation of kynurenine excretion with liver tryptophan pyrrolase levels in disease and after hydrocortisone induction. *Journal of Clinical Investigation, 45:* 1527-1534.

Andén, N.E., Rubenson, A., Fuxe, K., and Hökfelt, T. (1967): Evidence for dopamine receptor stimulation by apomorphine. *Journal of Pharmacy and Pharmacology, 19:* 627-629.

Andén, N.E., Corrodi, H., Fuxe, K., and Hökfelt, T. (1968): Evidence for a central 5-hydroxytryptamine receptor stimulation by lysergic acid diethylamine. *British Journal of Pharmacology, 34:* 1-7.

Andén, N.E., Butcher, S.G., Corrodi, H., Fuxe, K., and Ungerstedt, U. (1970): Receptor activity and turnover of dopamine and noradrenaline after neuroleptics. *European Journal of Pharmacology, 11:* 303-314.

Andén, N.E., Corrodi, H., Fuxe, K., and Hökfelt, T. (1968): Evidence for a central 5-catecholamine turnover assessed using tyrosine- and dopamine-β-hydroxylase inhibitors. *Journal of Pharmacy and Pharmacology, 24:* 177-182.

Andén, N.E., Engel, J., and Rubensson, A. (1972b): Central decarboxylation and uptake of l-dopa. *Naunyn-Schmiedeberg's Archives of Pharmacology, 273:* 11-26.

Andersson, H., and Roos, B-E. (1968): 5-Hydroxyindoleacetic acid in ventricular cerebrospinal fluid and brain of normal and hydrocephalic dogs after administration of 5-hydroxytryptophan. *Acta Pharmacologica (Kbh)., 26:* 531-538.

Andersson, H. and Roos, B-E. (1969): 5-Hydroxyindoleacetic acid in cerebrospinal fluid of hydrocephalic children. *Acta Paediatrica* (Uppsala), *58:* 601-608.

135

Angel, Ch., Deluca, D.C., and Murphree, O.D. (1976): Effects of drugs and stress on genetically nervous dogs. *Biological Psychiatry* (In press).

Angst, J. (1966): Zur Ätiologie und Nosologie endogener depressiver Psychosen: eine genetische, soziologische und klinische Studie. Springer, Berlin.

Angst, J., Baastrup, P., Grof, P., Hippius, H., Pöldinger, W., and Weis, P. (1973): The course of monopolar depression and bipolar psychoses. *Psychiatrica, Neurologia, Neurochirurgia* (Amsterdam), *76:* 489-500.

Anton-Tay, F., and Wurtman, R.J. (1971): Brain monoamines and endocrine function. In: *Frontiers in Neuroendocrinology.* Edited by L. Martini and W.F. Ganong. Oxford University Press, New York. Pp. 45-66.

Åsberg, M., Bertilsson, L., Tuck, D., Cronholm, B., and Sjöqvist, F. (1972): Indolamine metabolites in the cerebrospinal fluid of depressed patients before and during treatment with nortriptyline. *Clinical Pharmacology and Therapeutics, 14:* 277-286.

Ashcroft, G.W., and Sharman, D.F. (1960): 5-Hydroxyindoles in human cerebrospinal fluids. *Nature, 186:* 1050-1051.

Ashcroft, G.W., Eccleston, D., Knight, F., McDougall, E.J., and Waddell, J.L. (1965): Changes in amine metabolism produced by antidepressant drugs. *Journal of Psychosomatic Research, 9:* 129-136.

Ashcroft, G.W., Crawford, T.B.B., Eccleston, D., Sharman, D.F., McDougall, E.J., Stanton, J.B., and Binns, J.K. (1966): 5-Hydroxyindole compounds in the cerebrospinal fluid of patients with psychiatric or neurological diseases. *Lancet, II:* 1049-1052.

Ashcroft, G.W., Dow, R.C., and Moir, A.T.B. (1968): The active transport of 5-hydroxyindole-3-glacetic acid and 3-methoxy-4-hydroxyphenylacetic acid from a recirculatory perfusion system of the cerebral ventricles of the unanaesthetized dog. *Journal of Physiology, 199:* 397-424.

Ashcroft, G.W., Crawford, T.B.B., Dow, R.C., and Moir, A.T.B. (1969): Release of amine metabolite into ventricular perfusion fluid as an index of turnover. In: *Metabolism of Amines in the Brain.* Edited by G. Hooper. Macmillian, London. Pp. 65-69.

Ashcroft, G.W., Eccleston, D., Murray, L.G., Glen, A.I.M., Crawford, T.B.B., Pullar, I.A., Shields, P.J., Walter, D.S., Blackburn, I.M., Connechan, J., and Lonergan, M. (1972): Modified amine hypothesis for the aetiology of affective illness. *Lancet, II:* 573-577.

Ashcroft, G.W., Crawford, T.B.B., Cundall, R.L., Davidson, D.L., Dobson, J., Dow, R.C., Eccleston, D., Loose, R.W., and Pullar, I.A. (1973a): 5-Hydroxytryptamine metabolism in affective illness: the effect of tryptophan administration. *Psychological Medicine, 3:* 326-332.

Ashcroft, G.W., Blackburn, I.M., Eccleston, D., Glen, A.I.M., Hartley, W., Kinloch, N.E., Lonergan, M., Murray, L.G., and Pullar, I.A. (1973b): Changes on recovery in the concentrations of tryptophan and the biogenic amine metabolites in the cerebrospinal fluid of patients with affective illness. *Psychological Medicine, 3:* 319-325.

Axelrod, J., Whitby, L.G., and Hertting, G. (1961): Effect of psychotropic drugs on the uptake of H^3-norepinephrine by tissue. *Science, 133:* 383-384.

Axelrod, J. (1966): Methylation reactions in the formation and metabolism of catecholamines and other biogenic amines. *Pharmacological Reviews, 18:* 95-113.

Axelrod, J., and Cohn, C.K. (1971): Methyl transferase enzymes in red blood cells. *Journal of Pharmacology and Experimental Therapeutics, 176:* 650-654.

Axelrod, J. (1972): Dopamine-β-hydroxylase: regulation of its synthesis and release from nerve terminals. *Pharmacological Reviews, 24:* 233-243.

Bachelard, H.S. (1974): *Brain Biochemistry.* Chapman & Hall Ltd., London.

Baldessarini, R.J., and Yorke, C. (1970): Effects of lithium and of pH on synaptosomal metabolism of noradrenaline. *Nature, 228:* 1301-1303.

Barkai, A., Glusman, M., and Rapport, M.M. (1972): Serotonin turnover in the intact cat brain. *Journal of Pharmacology and Experimental Therapeutics, 181:* 28-35.

Bartholini, G., Pletscher, A., and Tissot, R. (1966): On the origin of homovanillic acid in the cerebrospinal fluid. *Experientia* (Basel), *22:* 609-610.

Bartholini, G., and Pletscher, A. (1968): Cerebral accumulation and metabolism of C^{14}-dopa after selective inhibition of peripheral decarboxylase. *Journal of Pharmacology and Experimental Therapeutics, 161:* 14-20.

Becker, J. (1974): *Depression: Theory and Research.* Winston, Washington.

Beckmann, H., and Goodwin, F.K. (1975): Antidepressant response to tricyclics and urinary MHPG in unipolar patients. *Archives of General Psychiatry, 32:* 17-21.

Beckmann, H., St.-Laurent, J., and Goodwin, F.K. (1975): The effect of lithium on urinary MHPG in unipolar and bipolar depressed patients. *Psychopharmacologia* (Berl.), *42,* 277-282.

Benkert, O., Renz, A., Marano, C., and Matussek, N. (1971): Altered tyrosine daytime plasma levels in endogenous depressive patients. *Archives of General Psychiatry, 25:* 359-363.

Benkert, O., Martschke, D., and Gordon, A. (1974): Comparison of TRH, L.H.-R.H., and placebo in depression. *Lancet, II:* 1146.

Berg, G.R., and Glinsmann, W.H. (1970): Cyclic AMP in depression and mania. *Lancet, I:* 834.

Bernheimer, H., and Hornykiewicz, O. (1964): Das Verhalten des Dopamine Metaboliten Homovanillinsäure im Gehirn von Normalen und Parkinson-kranken Menschen. *Archiv für experimentelle Pathologie und Pharmakologie, 247:* 305-306.

Bernheimer, H., Birkmayer, W., and Hornykiewicz, O. (1966): Homovanillinsäure in Liquor cerebrospinalis: Untersuchungen beim Parkinson-Syndrom und anderen Erkrankungen des ZNS. *Wiener klinische Wochenschrift, 78:* 417-419.

Bertaccini, G. (1959): Effect of convulsant treatment on the 5-hydroxytryptamine content of brain and other tissues of the rat. *Journal of Neurochemistry, 4:* 217-222.

Bertilsson, L., Åsberg, M. and Thorén, P. (1974): Differential effect of chlorimipramine and nortriptyline on metabolites of serotonine and noradrenaline in the cerebrospinal fluid of depressed patients. *European Journal of Clinical Pharmacology, 7:* 365-368.

Bertilsson, L., and Åsberg, M. (1975): Determination of biogenic amine metabolites in CSF by mass fragmentography. Methods and biochemical studies of depressive disorders. Paper read at the Sixth International Congress of Pharmacology, Helsinki (in press).

Beskow, J., Gottfries, C.G., Roos, B.E., and Winblad, B. (1976): Determination of monoamine and monoamine metabolites in the human brain: post mortem studies in a group of suicides and in a control group. *Acta Psychiatrica Scandinavica, 53:* 7-20.

Bevan-Jones, B., Pare, C.M.B., Nicholson, W.J.N., Price, K., and Stacey, R.S. (1972): Brain amine concentrations after monoamine oxidase inhibitor administration. *British Medical Journal, 1:* 17-19.

Bindler, E.H., Wallach, M.B., and Gershon, S. (1971): Effect of lithium on the release of ^{14}C-norepinephrine by nerve stimulation from the perfused cat spleen. *Archives Internationales de Pharmacodynamie et de Thérapie, 190:* 150-154.

Birkmayer, W., and Hornykiewicz, O. (1962): Der L-dioxyphenylalanin (= L Dopa)-effekt beim Parkinson-Syndrom des Menschen: zur Pathogenese und Behandlung der Parkinson-Akinese. *Archiv für Psychiatrie und Nervenkrankheiten, 203:* 560-574.

Birkmayer, W., and Linauer, W. (1970): Störung des Tyrosin- und Tryptophanmetabolismus bei Depression. *Archiv für Psychiatrie und Nervenkrankheiten, 213:* 377-387.

Blackburn, K.J., French, P.C., and Merrils, R.J. (1967): 5-Hydroxytryptamine uptake in rat brain in vitro. *Life Sciences, 6:* 1653-1663.

Blaschko, H., Burn, J.H., and Langemann, H. (1950): The formation of noradrenaline from dihydroxyphenylderine. *British Journal of Pharmacology, 5:* 431-437.

Bliss, E.L., Thatcher, W., and Ailion, J. (1972): Relationship of stress to brain serotonin and 5-hydroxyindoleacetic acid. *Journal of Psychiatric Research, 9:* 71-80.

Bockar, J., Roth, R., and Heninger, G. (1974): Increased human platelet monoamine oxidase activity during lithium carbonate therapy. *Life Sciences, 15:* 2109-2118.

Bodganski, D.F.A., Pletscher, A., Brodie, B., and Udenfriend, S. (1956): Identification and assay of serotonin in brain. *Journal of Pharmacology and Experimental Therapeutics, 117:* 82-88.

Bohacek, N. (1973): Depression-inducing potential of neuroleptics. In: *Aspects of Depression.* Edited by M. Lader and R. García. World Psychiatric Association, Madrid. Pp. 225-231.

Bond, P.A., Jenner, F.A., and Sampson, G.A. (1972): Daily variations of the urine content of 3-methoxy-4-hydroxyphenylglycol in two manic-depressive patients. *Psychological Medicine, 2:* 81-85.

Bowers, M.B., Jr., and Gerbode, F. (1968a): Relationship of monoamine metabolites in human cerebrospinal fluid to age. *Nature, 219:* 1256-1257.

Bowers, M.B., Jr., and Gerbode, F. (1968b): CSF 5-HIAA. Effects of probenecid and parachlorophenylalanine. *Life Sciences, 7:* 773-776.

Bowers, M.B., Jr. (1969): Deficient transport mechanism for the removal of acid monoamine metabolites from cerebrospinal fluid. *Brain Research, 15:* 522-524.

Bowers, M.B., Jr., Heninger, G.R., and Gerbode, F. (1969): Cerebrospinal fluid 5-hydroxyindoleacetic acid and homovanillic acid in psychiatric patients. *International Journal of Neuropharmacology, 8:* 255-262.

Bowers, M.B., Jr. (1972a): Cerebrospinal fluid 5-hydroxyindoleacetic acid (5-HIAA) and homovanillic acid (HVA) following probenecid in unipolar depressives treated with amitriptyline. *Psychopharmacologia* (Berl), *23:* 26-33.

Bowers, M.B., Jr. (1972b): Acute psychosis induced by psychomimetic drug abuse: II. Neurochemical findings. *Archives of General Psychiatry, 27:* 440-442.

Bowers, M.B., Jr. (1972c): Clinical measurements of central dopamine and 5-hydroxytryptamine metabolism: reliability and interpretation of cerebrospinal fluid acid monoamine metabolite measures. *Neuropharmacology, 11:* 101-111.

Bowers, M.B., Jr. (1973): 5-Hydroxyindoleacetic acid (5-HIAA) and homovanillic acid (HVA) following probenecid in acute psychotic patients treated with phenotiazines. *Psychopharmacologia, 28:* 309.

Bowers, M.B., Jr. (1974): Amitriptyline in man: decreased formation of central 5-hydroxyindoleacetic acid. *Clinical Pharmacology and Therapeutics, 15:* 167-170.

Bourne, H.R., Bunney, W.E., Jr., Colburn, R.W., Davis, J.M., Davis, J.N., Shaw, D.M., and Coppen, A.J. (1968): Noradrenaline, 5-hydroxytryptamine and 5-hydroxyindoleacetic acid in hindbrain of suicidal patients. *Lancet, II:* 805-808.

Boyland, E. (1958): The biochemistry of cancer of the bladder. *British Medical Bulletin, 14:* 153-158.

British Medical Journal (1966): Methyldopa in hypertension. *1:* 119-120.

Brittain, R.T. (1966): The intracerebral effects of noradrenaline and its modification by drugs in the mouse. *Journal of Pharmacy and Pharmacology, 18:* 621-623.

Brodie, H.K.H., Murphy, D.L., Goodwin, F.K., and Bunney, W.E., Jr. (1971): Catecholamines and mania: the effect of alpha-methyl-para-tyrosine on manic behavior and catecholamine metabolism. *Clinical Pharmacology and Therapeutics, 12:* 219-224.

Brodie, H.K.H., Sack, R., and Siever, L. (1973): Clinical studies of 5-hydroxytryptophan in depression. In: *Serotonin and Behavior.* Edited by J. Barchas and E. Usdin. Academic Press, New York. Pp. 549-559.

Brown, G.W., Sklair, F., Harris, T.O., and Birley, J.L.T. (1973): Life-events and psychiatric disorder. Part I: Some methodological issues. *Psychological Medicine, 3:* 74-87.

Bruinvels, J. (1972): Inhibition of the biosynthesis of 5-hydroxytryptamine in rat brain by imipramine. *European Journal of Pharmacology, 20:* 231-237.

Bulat, M., and Zivkovic, B. (1971): Origin of 5-hydroxyindoleacetic acid in the spinal fluid. *Science, 173:* 738-740.

Bulat, M., and Zivkovic, B. (1973): Penetration of 5-hydroxyindoleacetic acid across the blood-cerebrospinal fluid barrier. *Journal of Pharmacy and Pharmacology, 25:* 178-179.

Bullpitt, C.J., and Dollery, C.T. (1973): Side effects of hypotensive agents evaluated by a self-administered questionnaire. *British Medical Journal, 3:* 485-490.

Bunney, B.S., Walters, J.R., Roth, R.H., and Aghajanian, G.K. (1973): Dopaminergic neurons: effect of antipsychotic drugs and amphetamine on single cell activity. *Journal of Pharmacology and Experimental Therapeutics, 185:* 560-571.

Bunney, W.E., Jr., and Davis, J.M. (1965): Norepinephrine in depressive reactions: review. *Archives of General Psychiatry, 13:* 483-494.

Bunney, W.E., Jr., Murphy, D.L., and Goodwin, F.K. (1970): The switch process from depression to mania: relationship to drugs which alter brain amines. *Lancet, I:* 1022-1027.

Bunney, W.E., Jr., Brodie, H.K.H., Murphy, D.L., and Goodwin, F.K. (1971): Studies of alpha-methyl-para-tyrosine, L-dopa and L-tryptophan in depression and mania. *American Journal of Psychiatry, 127:* 872-881.

Burg, W. van den, Praag, H.M. van, Bos, E.R.H., Zanten, A.K. van, Piers, D.A., and Doorenbos, H. (1975a): TRH as a possible quick-acting but short-lasting antidepressant. *Psychological Medicine, 5:* 404-412

Burg, W. van den, Praag, H.M. van, Bos, E.R.H., Zanten, A.K. van, Piers, D.A., and Doorenbos, H. (1975b): TRH by slow, continuous infusion: an antidepressant? *Psychological Medicine,* (in press).

Butcher, L.L., and Engel, J. (1969): Behavioural and biochemical effects of L-dopa after peripheral decarboxylase inhibition. *Brain Research, 15:* 233-242.

Cairncross, K.D. (1965): On the peripheral pharmacology of amitriptyline. *Archives Internationales de Pharmacodynamie et de Thérapie, 154:* 438-448.

Calne, D.B., Karoum, F., Ruthven, C.R.J., and Sandler, M. (1969): The metabolism of orally administered L-dopa in parkinsonism. *British Journal of Pharmacology, 37:* 57-68.

Campanini, T., Catalano, A., Derisio, C., and Mardighian, G. (1970): Vanilmandelic aciduria in the different clinical phases of manic depressive psychoses. *British Journal of Psychiatry, 116:* 435-436.

Carlsson, A., Rosengren, E., Bertler, A., and Nilsson, J. (1957): Effect of reserpine on the metabolism of catecholamines. In: *Psychotropic Drugs.* Edited by S. Garattini and V. Ghetti. Elsevier, Amsterdam. Pp. 363-372.

Carlsson, A., Lindqvist, M., and Magnusson, T. (1957): 3, 4-Dihydroxyphenylalanine and 5-hydroxytryptophan as reserpine antagonists. *Nature, 180:* 1200.

Carlsson, A., Falck, B., Fuxe, K., and Hillarp, N.A. (1964): Cellular localization of monoamines in the spinal cord. *Acta Physiology Scandinavica, 60:* 112-119.

Carlsson, A., Fuxe, K., Hamberger, B., and Lindqvist, M. (1966): Biochemical and histochemical studies on effects of imipramine-like drugs and (+)amphetamine on central and peripheral catecholamine neurons. *Acta Physiology Scandinavica, 67:* 481-497.

Carlsson, A., Fuxe, K., and Ungerstedt, U. (1968): Effect of imipramine on central 5-hydroxytryptamine neurons. *Journal of Pharmacy and Pharmacology, 20:* 150.

Carlsson, A., Corrodi, H., Fuxe, K., and Hökfelt, T. (1969a): Effect of antidepressant drugs on the depletion of intraneuronal brain 5-hydroxytryptamine stores caused by 4-methyl-α-ethyl-meta-tyramine. *European Journal of Pharmacology, 5:* 357-366.

Carlsson, A., Corrodi, H., Fuxe, K., and Hökfelt, T. (1969b): Effects of some antidepressant drugs on the depletion of intraneuronal brain catecholamine stores caused by 4, α-dimethyl-meta-tyramine. *European Journal of Pharmacology, 5:* 367-373.

Carlsson, A., Jonason, J., and Lindqvist, M. (1969c): On the mechanism of 5-hydroxytryptamine release by thymoleptics. *Journal of Pharmacy and Pharmacology, 21:* 769-773.

Carlsson, A., Kehr, W., Lindqvist, M., Magnusson, T., and Atack, C.V. (1972): Regulation of monoamine metabolism in the central nervous system. *Pharmacological Reviews, 24:* 371-384.

Carman, J.S., Post, R.M., Teplitz, T.A., and Goodwin, F.K. (1974): Divalent cautions in predicting antidepressant response to lithium. *Lancet, II:* 1454.

Carney, M.W.P., Thakurdas, H., and Sebastian, J. (1969): Effects of imipramine and reserpine in depression. *Psychopharmacologia, 14:* 349-350.

Carney, M.W.P., and Sheffield, B.F. (1974): The effects of pulse ECT in neurotic and endogenous depression. *British Journal of Psychiatry, 125:* 91-94.

Carroll, B.J., Mowbray, R.M., and Davies, B.M. (1970): Sequential comparison of l-tryptophan with E.C.T. in severe depression. *Lancet, I:* 967-969.

Carroll, B.J. (1971): Monoamine precursors in the treatment of depression. *Clinical Pharmacology and Therapeutics, 12:* 743-761.

Carroll, B.J., and Dodge, J. (1971): L-tryptophan as an antidepressant. *Lancet, I:* 915.

Cazzullo, C.L., Mangoni, A., and Mascherpa, G. (1966): Tryptophan metabolism in affective psychoses. *British Journal of Psychiatry, 112:* 157-162.

Charalampous, K.D., and Brown, S. (1967): A clinical trial of α-methyl-para-tyrosine in mentally ill patients. *Psychopharmacologia,* (Berl.), *11:* 422-429.

Chase, T.N. (1972): Serotonergic mechanisms in Parkinson's disease. *Archives of Neurology* (Chic.), *27:* 354-356.

Chase, T.N., Gordon, E.K., and Ng, L.K.Y. (1973): Norepinephrine metabolism in the central nervous system of man: studies using 3-methoxy-4-hydroxyphenylethylene glycol levels in cerebrospinal fluid. *Journal of Neurochemistry, 21:* 581-587.

Chodoff, P. (1972): The depressive personality. A critical review. *Archives of General Psychiatry, 27:* 666-673.

Claveria, L.E., Curzon, G., Harrison, M.J.G., and Kantamaneni, B.D. (1974): Amine metabolites in the cerebrospinal fluid of patients with disseminated sclerosis. *Journal of Neurology, Neurosurgery and Psychiatry, 37:* 715-718.

Cohn, C.K., Dunner, D.L., and Axelrod, J. (1970): Reduced catechol-O-methyltransferase activity in red blood cells of women with primary affective disorder. *Science, 170:* 1323-1324.

Consolo, S., Dolfini, E., Garattini, S., and Valzelli, L. (1967): Desipramine and amphetamine metabolism. *Journal of Pharmacy and Pharmacology, 19:* 253-256.

Constantinides, J., Geissbühler, F., Gaillard, J.M., Hovaguimian, Th., and Tissot, R. (1974): Enhancement of cerebral noradrenaline turnover by thyrotropin-releasing hormone: evidence by fluorescence histochemistry. *Experientia, 30:* 1182.

Cooper, J.R., Bloom, F.E., and Roth, R.H. (1970): *The Biochemical Basis of Neuropharmacology.* Oxford University Press, New York.

Coppen, A., Shaw, D.M., and Farrell, J.P. (1963): Potentiation of the antidepressive effects of a monoamine oxidase inhibitor by tryptophan. *Lancet, I:* 79-81.

Coppen, A., Shaw, D.M., and Malleson, A. (1965a): Changes in 5-hydroxytryptophan metabolism in depression. *British Journal of Psychiatry, 111:* 105-107.

Coppen, A., Shaw, D.M., Malleson, A., Eccleston, E., and Gundy, G. (1965b): Tryptamine metabolism in depression. *British Journal of Psychiatry, 111:* 996-998.

Coppen, A. (1967): Biochemistry of affective disorders. *British Journal of Psychiatry, 113:* 1237-1264.

Coppen, A., Shaw, D.M., Herzberg, B., and Maggs, R. (1967): Tryptophan in treatment of depression. *Lancet, II:* 1178-1180.

Coppen, A., Prange, A.J., Whybrow, P.C., Noguera, R., and Paez, J.M. (1969): Methysergide in mania. *Lancet, II:* 338-340.

Coppen, A., Prange, A.J., Whybrow, P.C., and Noguera, R. (1972a): Abnormalities of indolamines in affective disorders. *Archives of General Psychiatry, 26:* 474-478.

Coppen, A., Brooksbank, B.W.L., and Peet, M. (1972b): Tryptophan concentration in the cerebrospinal fluid of depressive patients. *Lancet, I:* 1393.

Coppen, A., Eccleston, E.G., and Peet, M. (1973): Total and free tryptophan concentration in the plasma of depressive patients. *Lancet, II:* 60-63

Coppen, A., Brooksbank, B.W.G., and Eccleston, E. (1974a): Tryptophan metabolism in depressive illness. *Psychological Medicine, 4:* 164-173.

Coppen, A., Montgomery, S., Peet, M., Baily, J., Marks, V., and Woods, P. (1974b): Thyrotropin-releasing hormone in the treatment of depression. *Lancet, II:* 433-435.

Corrodi, H., Fuxe, K., Hökfelt, T., and Schou, M. (1967): Effect of lithium on cerebral monoamine neurons. *Psychopharmacologia, 11:* 345-353.

Corrodi, H., and Fuxe, K. (1968): The effect of imipramine on central monoamine neurons. *Journal of Pharmacy and Pharmacology, 20:* 230-231.

Corrodi, H., and Fuxe, K. (1969): Decreased turnover in central 5-HT nerve terminals induced by antidepressant drugs of the imipramine type. *European Journal of Pharmacology, 7:* 56-59.

Corrodi, H., Fuxe, K., and Schou, M. (1969): The effect of prolonged lithium administration on cerebral monoamine neurons in the rat. *Life Sciences, 8:* 643-651.

Cotzias, M., Woert, M.H. van, and Schiffer, L.M. (1967): Aromatic aminoacids and modifications of parkinson. *New England Journal of Medicine, 276:* 374-379.

Cramer, H., Goodwin, F.K., Post, R.M., and Bunney, W.E., Jr. (1972): Effects of probenecid and exercise on cerebrospinal fluid cyclic AMP in affective illness. *Lancet, I:* 1346-1347.

Cremata, V.Y., and Koe, B.K. (1966): Clinical-pharmacological evaluation of p-chlorophenylalanine: new serotonin-depleting agent. *Clinical Pharmacology and Therapeutics, 7:* 768-776.

Creveling, C.R., Daly, J., Tokuyama, T., and Witkop, B. (1968): The combined use of α-methyltyrosine and threo-dihydroxy-phenylserine—selective reduction of dopamine levels in the central nervous system. *Biochemical Pharmacology, 17:* 65-70.

Crout, J.R., and Sjoerdsma, A. (1959): Clinical and laboratory significance of serotonin and catecholamines in bananas. *New England Journal of Medicine, 261:* 23-26.

Curzon, G., and Green, A.R. (1968): Effect of hydrocortisone on rat brain 5-hydroxytryptamine. *Life Sciences, 7:* 657-663.

Curzon, G. (1969): Tryptophan pyrrolase—a biochemical factor in depressive illness? *British Journal of Psychiatry, 115:* 1367-1374.

Curzon, G., Gumpert, E.J.W., and Sharpe, D.M. (1971): Amine metabolites in the lumbar cerebrospinal fluid of humans with restricted flow of cerebrospinal fluid. *Nature/New Biology, 231:* 189-191.

Curzon, G., Friedel, J., and Knott, P.J. (1973): The effect of fatty acids on the binding of tryptophan to plasma protein. *Nature, 242:* 198-200.

Dahlström, A., Haggendal, J., and Atack, C. (1973): Localization and transport of serotonin. In: *Serotonin and Behavior.* Edited by J. Barchas and E. Usdin. Academic Press, New York. Pp. 87-96.

Davson, H. (1967): *Physiology of the Cerebrospinal Fluid*. Churchill, London.

Degkwitz, R., Frowein, R., Kulenkampff, C., and Mohs, V. (1960): Uber die Wirkungen des L-Dopa beim Menschen und deren Beeinflussung durch Reserpin, Chlorpromazin, Iproniazid, und Vitamin B₆. *Klinische Wochenschrift, 38:* 120-123.

Deleon-Jones, F., Maas, J.W., Dekirmenjian, H., and Sanchez, J. (1975): Diagnostic subgroups of affective disorders and their urinary excretion of catecholamine metabolites. *American Journal of Psychiatry, 132,* 1141-1148.

Dengles, H.G., and Titus, E.O. (1961): The effect of drugs on the uptake of isotopic norepinephrine in various tissues. *Biochemical Pharmacology, 8:* 64.

Denker, S.J., Malm, U., Roos, B-E., and Werdinius, B. (1966): Acid monoamine metabolites of cerebrospinal fluid in mental depression and mania. *Journal of Neurochemistry, 13:* 1545-1548.

Despopoulos, A., and Weissbach, H. (1957): Renal metabolism of 5-hydroxyindoleacetic acid. *American Journal of Physiology, 189:* 548-550.

Dewhurst, W.G. (1968): Methysergide in mania. *Nature, 219:* 506-507.

Doepfner, W., and Cerletti, A.M.S. (1958): Comparison of lysergic acid derivatives and antihistamines as inhibitors of the edema provoked in the rat's paw by serotonin. *International Archives of Allergy, 12:* 89-97.

Doepfner, W. (1962): Biochemical observations on LSD-25 and Deseril. *Experientia* (Basle), *18:* 256-257.

Domenjoz, R., and Theobald, W. (1959): Zur Pharmakologie des Tofranil (N-(3-dimethyl-aminopropyl)-iminodibenzylhydrochlorid). *Archives Internationales de Pharmacodynamie et de Thérapie, 120:* 450-489.

Dousa, T., and Hechter, O. (1970): Lithium and brain adenyl cyclase. *Lancet, I:* 834-835.

Dunlop, E., DeFelice, E.A., Bergen, J.R., and Resnick, O. (1965): The relationship between MAO inhibition and improvement of depression: preliminary results with intravenous modaline sulfate (W3207B). *Psychosomatics, 6:* 1-7.

Dunner, D.L., Cohn, C.K., Gershon, E.S., and Goodwin, F.K. (1971): Differential catechol-O-methyltransferase activity in unipolar and bipolar affective illness. *Archives of General Psychiatry, 25:* 348-353.

Dunner, D.L., and Goodwin, F.K. (1972): Effect of l-tryptophan on brain serotonin metabolism in depressed patients. *Archives of General Psychiatry, 26:* 364-366.

Dunner, D.L., and Fieve, R.R. (1975): Affective disorders. Studies with amine precursors. *American Journal of Psychiatry, 132:* 180-183.

Duvoisin, R.C., Yahr, M.D., and Cote, L.D. (1969): Pyridoxine reversal of L-dopa effects in parkinsonism. *Transactions of the American Neurological Association, 94:* 81-84.

Ebert, M.H., Baldessarini, R.J., Lipinski, J.F., and Berv, K. (1973): Effects of electro-convulsive seizures on amine metabolism in the rat brain. *Archives of General Psychiatry, 29:* 397-401.

Eble, J.N. (1964): A study of the mechanisms of the modifying actions of cocaine, ephedrine and imipramine on the cardiovascular response to norepinephrine and epinephrine. *Journal of Pharmacology and Experimental Therapeutics, 144:* 76-82.

Eccleston, D., Ashcroft, G.W., Moir, A.T.B., Parker-Rhodes, A., Lutz, W., and O'Mahoney, D.P. (1968): A comparison of 5-hydroxyindoles in various regions of dog brain and cerebrospinal fluid. *Journal of Neurochemistry, 15:* 947-957.

Eccleston, D., Loose, R., Pullar, J.A., and Sugden, R.F. (1970a): Exercise and urinary excretion of cyclic AMP. *Lancet, II:* 612-613.

Eccleston, D., Ashcroft, G.W., Crawford, T.B.B., Stanton, J.B., Wood, D., and McTurk, P.H. (1970b): Effect of tryptophan administration on 5-HIAA in cerebrospinal fluid in man. *Journal of Neurology, Neurosurgery and Psychiatry, 33:* 269-272.

Edwards, D.J., and Burns, M.O. (1974): Effects of tricyclic antidepressants upon human platelet monoamine oxidase. *Life Sciences, 15:* 2045-2058.

Ehringer, H., and Hornykiewicz, O. (1960): Verteilung von Noradrenalin und Dopamin (3-hydroxytryptamine) im Gehirn des Menschen und Ihr Verhalten bei Erkrankungen des extrapyramidalen Systems. *Klinische Wochenschrift, 38:* 1236-1239.

Eleftherion, B.E., and Boehlke, K.W. (1967): Brain monoamine oxidase in mice after exposure to agression and defeat. *Science, 155:* 1693-1694.

El-Yousef, M.K., Janowski, D.S., Davis, J.M., and Sekerke, H.J. (1973): Induction of severe depression by physostigmine in marijuana intoxicated individuals. *British Journal of Addiction, 68:* 321-326.

Engel, J., Hanson, L.C.F., and Roos, B-E. (1971): Effect of electroshock on 5-HT metabolism in rat brain. *Psychopharmacologia, 20:* 197-200.

Engelman, K., Lovenberg, W., and Sjoerdsma, A. (1967): Inhibition of serotonin synthesis by para-chlorophenylalanine in patients with the carcinoid syndrome. *New England Journal of Medicine, 277:* 1103-1108.

Erspamer, V. (1966): Occurrence of indolealkylamines in nature. In: *5-Hydroxytryptamine and Related Indolealkylamines.* Edited by V. Erspamer. Springer, Berlin-Heidelberg-New York. Pp. 132-181.

Everett, G.M., and Borcherding, J.W. (1970): L-dopa: effect on concentrations of dopamine, norepinephrine and serotonin in brains of mice. *Science, 168:* 849-850.

Extein, I., Korf, J., Roth, R.H., and Bowers, M.B. Jr. (1973): Accumulation of 3-methoxy-4-hydroxyphenylglycol-sulfate in rabbit cerebrospinal fluid following probenecid. *Brain Research, 54:* 403-407.

Fann, W.E., Davis, J.M., Janowski, D.S., Cavanaugh, J.H., Kaufmann, J.S., Griffith, J.D., and Oates, J.A. (1972): Effects of lithium on adrenergic function in man. *Clinical Pharmacology and Therapeutics, 13:* 71-77.

Farnebo, L.O., and Hamberger, B. (1971): Drug-induced changes in the release of ³H-noradrenaline from field stimulated rat iris. *British Journal of Pharmacology, 43:* 97-106.

Fernando, J.C., Joseph, M.H., and Curzon, G. (1975): Tryptophan plus a pyrrolase inhibitor for depression? *Lancet, I:* 171.

Fernström, J.D., and Wurtman, R.J. (1971): Brain serotonin content: physiological dependence on plasma tryptophan levels. *Science, 173:* 149-152.

Fernström, J.D., and Wurtman, R.J. (1972). Brain serotonin content: physiological regulation by plasma neutral amino acids. *Science, 178:* 414-416.

Fieve, R.R., Platman, S.R., and Fliess, J.L. (1969): A clinical trial of methysergide and lithium in mania. *Psychopharmacologia, 15:* 425-429.

Fischbach, R., Harrer, G., and Harrer, H. (1966): Verstärkung der Noradrenaline Wirkung durch Psychopharmaka beim Menschen. *Arzneimittelforschung, 2(a):* 263-265.

Fischer, E., Heller, B., and Miro, A.N. (1968): β-phenylethylamine in human urine. *Arzneimittelforschung, 18:* 1486.

Fischer, E., Spatz, H., Saavedra, J.M., Reggiani, H., Miro, A.H., and Heller, B. (1972a): Urinary elimination of phenylethylamine. *Biological Psychiatry, 5:* 139-147.

Fischer, E., Spatz, H., Heller, B., and Reggiani, H. (1972b): Phenylethylamine content of human urine and rat brain: its alteration in pathological conditions and after drug administration. *Experientia, 28:* 307-308.

Flentge, F., and Klaver, M.M. (1975): An investigation into the origin of choline in cerebrospinal fluid. *Proceedings of the Federal Meetings, Utrecht, The Netherlands.*

Forn, J., and Valdescasas, F.G. (1971): Effects of lithium on brain adenyl cyclase activity. *Biochemical Pharmacology, 20:* 2773-2779.

Friedman, E., and Gershon, S. (1973): Effect of lithium on brain dopamine. *Nature, 243:* 520-521.

Frey, H.H., and Magnussen, M.P. (1968): Different central mediation of the stimulant effects of amphetamine and its p-chloro-analogue. *Biochemical Pharmacology, 17:* 1299-1308.

Fuller, R.W., Hines, C.W., and Mills, J. (1965): Lowering of brain serotonin level by chloro-amphetamines. *Biochemical Pharmacology, 14:* 483-488.

Fuller, R.W., Perry, K.W., and Molloy, B.B. (1974): Effect of an uptake inhibitor on serotonin metabolism in rat brain: studies with 3-(p-trifluoromethylphenoxy)-N-methyl-3-phenylpropylamine (Lilly 110140). *Life Sciences, 15:* 1161-1171.

Fuller, R.W., and Molloy, B.B. (1974): Recent studies with 4-chloroamphetamine and some analogues. *Advances in Biochemical Psychopharmacology, 10:* 195-205.

Fuxe, K., Hökfelt, T., Nilsson, O., and Reinus, S. (1966): A fluorescence and electron microscopic study on central monoamine nerve cells. *Anatomical Record, 155:* 33-40.

Fuxe, K., Butcher, L.L., and Engel, J. (1971): DL-5-hydroxytryptophan-induced changes in central monoamine neurons after peripheral decarboxylase inhibition. *Journal of Pharmacy and Pharmacology, 23:* 420-424.

Fyrö, B., Petterson, U., and Sedvall, G. (1975): The effect of lithium treatment on manic symptoms and levels of monoamine metabolites in cerebrospinal fluid of manic depressive patients. *Psychopharmacologia* (Berl.), *44,* 99-103.

Gabay, S., and Valcourt, A.J. (1968): Biochemical determinants in the evaluation of mono-amine-oxidase inhibitors. *Recent Advances in Biological Psychiatry, 10:* 29-41.

Garattini, S., Kato, R., Lamesta, L., and Valzelli, L. (1960): Electroshock, brain serotonin and barbiturate narcosis. *Experientia, 16:* 156-157.

Garelis, E., and Sourkes, T.L. (1973): Sites of origin in the central nervous system of mono-amine metabolites measured in human cerebrospinal fluid. *Journal of Neurology, Neurosurgery and Psychiatry, 4:* 625-629.

Garelis, E., Young, S.N., Lal, S., and Sourkes, T.L. (1974): Monoamine metabolites in lumbar CSF: the question of their origin in relation to clinical studies. *Brain Research, 79:* 1-8.

Garside, R.F., Kay, D.W.K., Wilson, I.C., Deaton, I.D., and Roth, M. (1971): Depressive syndromes and the classification of patients. *Psychological Medicine, 1:* 333-338.

Genefke, I.K. (1972): The concentration of 5-HT in hypothalamus, grey and white brain substance in the rat after prolonged oral lithium administration. *Acta Psychiatrica Scandinavica, 48:* 400-404.

Gershon, S., Hekimian, L.J., Floyd, A., Jr. and Hollister, L.E. (1967): α-methyl-p-tyrosine (AMT) in schizophrenia. *Psychopharmacologia, 11:* 189-194.

Gershon, S., Goodwin, F.K., and Gold, P. (1970): Effect of l-tyrosine and l-dopa on nor-epinephrine (NE) turnover in rat brain in vivo. *Pharmacologist, 12:* 268.

Gertner, S.B., Passonen, M.K., and Giarman, N.J. (1957): Presence of 5-hydroxytryptamine (serotonin) in perfusate from sympathetic ganglia. *Federation Proceedings, 16:* 299.

Glassman, A. (1969): Indoleamines and Affective disorders. *Psychosomatic Medicine, 31:* 107-114.

Glassman, A. and Platman, S.R. (1969): Potentiation of a monoamine oxidase inhibitor by tryptophan. *Journal of Psychiatric Research, 7:* 83-88.

Glowinski, J., and Axelrod, J. (1964). Inhibition of uptake of tritiated-noradrenaline in the intact rat brain by imipramine and structurally related compounds. *Nature, 204:* 1318-1319.

Glowinski, J., Kopin, I.J., and Axelrod, J. (1965): Metabolism of (^{3}H) norepinephrine in the rat brain. *Journal of Neurochemistry, 12:* 25-30.

Glowinski, J., and Axelrod, J. (1965): Effects of drugs on the uptake, release and metabolism of H^3-norepinephrine in the rat brain. *Journal of Pharmacology and Experimental Therapeutics, 149:* 43-49.

Glowinski, J., Iversen, L.L., and Axelrod, J. (1966a): Storage and synthesis of norepine-phrine in the reserpine-treated rat brain. *Journal of Pharmacology and Experimental Therapeutics, 151:* 385-399.

Glowinski, J., Axelrod, J., and Iversen, L.L. (1966b). Regional studies of catecholamines in the rat brain. *Journal of Pharmacology and Experimental Therapeutics, 153:* 30-41.

Glowinski, J., Besson, M.J., Cheramy, A., and Thierry, A.M. (1972a): Disposition and role of newly synthesized amines in central catecholaminergic neurons. In: *Advances in Biochemical Psychopharmacology.* Edited by E. Costa, L.L. Iversen and R. Paoletti. Raven Press, New York. Pp. 93-100.

Glowinski, J., Hamon, M., Javoy, O., and Morot-Gaudry, Y. (1972b): Rapid-effects of monoamine oxidase inhibitors on synthesis and release of central monoamines. *Advances in Biochemical Psychopharmacology, 5:* 423-439.

Godwin-Austen, R.B., Kantamaneni, B.D., and Curzon, G. (1971): Comparison of benefit from L-dopa in Parkinsonism with increase of amine metabolites in the CSF. *Journal of Neurology, Neurosurgery and Psychiatry, 34:* 219-223.

Goldstein, M., and Gerber, H. (1963): Phenolic alcohols in the brain after administration of dopa C^{14}. *Life Sciences, 2:* 97-100.

Goodwin, F.K., Brodie, H.K.H., Murphy, D.L., and Bunney, W.E., Jr. (1970): L-dopa, catecholamines and behavior: a clinical and biochemical study in depressed patients. *Biological Psychiatry, 2:* 341-366.

Goodwin, F.K., Dunner, D.L., and Gershon, S. (1971): Effect of l-dopa treatment on brain serotonin metabolism in depressed patients. *Life Sciences, 10:* 751-759.

Goodwin, F.K., Ebert, M.H., and Bunney, W.E., Jr. (1972a): Mental effects of reserpine in man: a review. In: *Psychiatric Complications of Medical Drugs.* Edited by R.I. Shader. Raven Press, New York. Pp. 73-101.

Goodwin, F.K. (1972): Behavioral effects of l-dopa in man. In: *Psychiatric Complications of Medical Drugs.* Edited by R.I. Shader. Raven Press, New York. Pp. 149-174.

Goodwin, F.K., Murphy, D.L., Dunner, D.L., and Bunney, W.E., Jr. (1972b): Lithium response in unipolar versus bipolar depression. *American Journal of Psychiatry, 129:* 44-47.

Goodwin, F.K., Post, R.M., Dunner, D.L., and Gordon, E.K. (1973): Cerebrospinal fluid amine metabolism in affective illness: the probenecid technique. *American Journal of Psychiatry, 130:* 73-79.

Goodwin, F.K., and Post, R.M. (1973): The use of probenecid in high doses for the estima-tion of central serotonin turnover in affective illness and addicts of methadone. In: *Serotonin and Behavior.* Edited by J. Barchas and E. Usdin. Academic Press, New York. Pp. 469-480.

Goodwin, F.K., and Post, R.M. (1974): Cerebrospinal fluid amine metabolites in affective illness. *Journal of Psychiatric Research, 10:* 320.

Goodwin, F.K. (1976): Discussion remarks. In: *Neuroregulators and Hypotheses of Psychi-atric Disorders.* Edited by J. Barchas, D.A. Hamburg, and E. Usdin. Oxford University Press. (in press)

Gordon, E.K., and Olivier, J. (1971): 3-Methoxy-4-hydroxy-phenyl-ethylene glycol in human cerebrospinal fluid. *Clinica Chimica Acta, 35:* 145-150.

Gordon, E.K., Olivier, J., Goodwin, F.K., Chase, T.N., and Post, R.M. (1973): Effect of probenecid on free 3-methoxy-4-hydroxyphenylethylene glycol (MHPG) and its sul-phate in human cerebrospinal fluid. *Neuropharmacology, 12:* 391-396.

Gottfries, C.G., and Roos, B-E. (1969): Homovanillic acid and 5-hydroxyindoleacetic acid in the cerebrospinal fluid of patients with senile dementia, presenile dementia and parkinsonism. *Journal of Neurochemistry, 16:* 1341-1345.

Gottfries, C.G., and Roos, B-E. (1970): Homovanillic acid and 5-hydroxy-indoleacetic acid in cerebrospinal fluid related to rated mental and motor impairment in senile and presenile dementia. *Acta Psychiatrica Scandinavica, 46:* 99-105.

Gottfries, C.G., and Roos, B-E. (1973): Acid monoamine metabolites in cerebrospinal fluid from patients with presenile dementia (Alzheimer's disease). *Acta Psychiatrica Scandinavica, 49:* 257-263.

Gottfries, C.G., Oreland, L., and Wiberg, A. (1974): Brain-levels of monoamine oxidase in depression. *Lancet, II:* 360.

Gottfries, C.G., Oreland, L., Wiberg, Å., and Winblad, B. (1975): Lowered monoamine oxidase activity in brains from alcoholic suicides. *Journal of Neurochemistry, 25,* 667-673.

Grabowska, M., Antkiewicz, L., and Michaluk, J. (1974): The influence of quipazine on the turnover rate of serotonin. *Biochemical Pharmacology, 23:* 3211-3212.

Grahame-Smith, D.G. (1971): Studies in vivo on the relationship between brain trypto-phan, brain 5-HT synthesis and hyperactivity in rats treated with a monoamine oxidase inhibitor and l-tryptophan. *Journal of Neurochemistry, 18:* 1053-1066.

Grahame-Smith, D.G., and Green, A.R. (1974): The role of brain 5-hydroxytryptamine in the hyperactivity produced in rats by lithium and monoamine oxidase inhibition. *British Journal of Pharmacology, 52:* 19-26.

Green, A.R., and Curzon, G. (1968): Decrease of 5-hydroxytryptamine in the brain pro-voked by hydrocortisone and its prevention by Allpurinol. *Nature, 220:* 1095-1097.

Green, A.R., and Grahame-Smith, D.G. (1974): TRH potentiates behavioural changes fol-lowing increased brain 5-hydroxy-tryptamine accumulation in rats. *Nature, 251:* 524-526.

Green, H., Greenberg, S.M., and Erickson, R.W. (1962): Effect of dietary phenylalanine and tryptophan upon the rat brain amine levels. *Journal of Pharmacology and Experimental Therapeutics, 136:* 174-178.

Greenspan, K., Schildkraut, J.J., Gordon, E.K., Levy, B., and Durell, J. (1969): Catechola-mine metabolism in affective disorders. II. Norepinephrine, normetanephrine, epine-phrine, metanephrine and VMA excretion in hypomanic patients. *Archives of General Psychiatry, 21:* 710-716.

Greenspan, K., Schildkraut, J.J., Gordon, E.K., Baer, L., Aranoff, M.S., and Durell, J. (1970): Catecholamine metabolism in affective disorders. III. MHPG and other cate-cholamine metabolites in patients treated with lithium carbonate. *Journal of Psychiatric Research, 7:* 171-183.

Grillo, M.A. (1970). Extracellular synaptic vesicles in the mouse heart. *Journal of Cell Biology, 47:* 547-553.

Grof, P., and Foley, P. (1971): The superiority of lithium over methysergide in treating manic patients. *American Journal of Psychiatry, 127:* 1573-1574.

Gruen, P.H., Sachar, E.J., Altman, N., and Sassin, J. (1975): Growth hormone responses to hypoglycemia in postmenopausal depressed women. *Archives of General Psychiatry, 32:* 31-33.

Guldberg, H.C., Ashcroft, G.W., and Crawford, T.B.B. (1966): Concentrations of 5-hy-droxyindoleacetic acid and homovanillic acid in the cerebrospinal fluid of the dog before and during treatment with probenecid. *Life Sciences, 5:* 1571-1575.

Gyermek, L., and Possemato, C. (1960): Potentiation of 5-hydroxytryptamine by imipra-mine. *Medicina Experimentalis, 3:* 225-229.

Haefely, von W., Hurlimann, A., and Thoenen, H. (1964): Scheinbar paradoxe Beeinflus-sung von peripheren Noradrenalin Wirkungen durch einige Thymoleptica. *Helvetia Physiologica und Pharmakologica Acta, 22:* 15-33.

Haggendal, J., and Lindqvist, M. (1964): Disclosure of labile monoamine fractions in the brain and their correlation to behavior. *Acta Physiologica Scandinavica, 60:* 351-357.

Halaris, A.E., Lovell, R.A., and Freedman, D.X. (1973): Effect of chloripramine on the metabolism of 5-hydroxytryptamine in the rat brain. *Biochemical Pharmacology, 22:* 2200-2202.

Halliwell, G., Quinton, R.M., and Williams, F.E. (1964): A comparison of imipramine, chlorpromazine and related drugs in various tests involving autonomic functions and antagonism of reserpine. *British Journal of Pharmacology, 23:* 330-350.

Hamadah, K., Holmes, H., Barker, G.B., Hartman, G.C., and Parke, D.V.W. (1972): Effect of electric convulsion therapy on urinary cyclic adenosine monophosphate. *British Medical Journal, III:* 439-441.

Hamberger, B., and Masuoka, D. (1965): Localization of catecholamine uptake in rat brain slices. *Acta Pharmacologica (Kbh), 22:* 363-368.

Hamilton, M. (1968): Some aspects of the long-term treatment of severe hypertension with methyldopa. *Postgraduate Medical Journal, 44:* 66-69.

Hamon, M., and Glowinski, J. (1974): Regulation of serotonin synthesis. *Life Sciences, 15:* 1533-1548.

Harper, A.E., Benevenga, N.J., and Wolhueter, R.M. (1970): Effects of ingestion of disproportionate amounts of amino acids. *Physiological Reviews, 50:* 428-558.

Haskovec, L., and Rysanek, K. (1967): The action of reserpine in imipramine-resistant depressive patients. *Psychopharmacologia, 11:* 18-30.

Haskovec, L., and Soucek, K. (1968): Trial of methysergide in mania. *Nature, 219:* 507-508.

Haskovec, L. (1969): Methysergide in mania. *Lancet, II:* 902.

Hassler, R., Bak, I.J., and Kim, J.S. (1970): Unterschiedliche Entleerung des Speicheorte Für Noradrenalin, Dopamine, und Serotonin als Wirkungsprinzip des Oxypertins. *Nervenarzt, 41:* 105-118.

Hendley, E., and Snyder, S.H. (1968): Relationship between the action of monoamine oxidase inhibitors on the noradrenaline uptake system and their antidepressant efficacy. *Nature, 220:* 1330-1331.

Hendley, E.D., and Welch, B.L. (1975): Electroconvulsive shock: sustained decrease in norepinephrine uptake affinity in a reserpine model of depression. *Life Sciences, 16:* 45-54.

Herrington, R.N., Bruce, A., Johnstone, E.C., and Lader, M.H. (1974): Comparative trial of l-tryptophan and E.C.T. in severe depressive illness. *Lancet, II:* 731-734.

Hertting, G., Axelrod, J., Kopin, I.J., and Whitby, L.G. (1961): Lack of uptake of catecholamines after chronic denervation of sympathetic nerves. *Nature, 189:* 66.

Hertting, G., and La Brosse, H. (1962): Bilary and urinary excretion of metabolites of 7 H^3-epinephrine in the rat. *Journal of Biological Chemistry, 237:* 2291-2295.

Hertz, D., and Sulman, F.G. (1968): Preventing depression with tryptophan. *Lancet, I:* 531-532.

Hinsley, R.K., Norton, J.A., and Aprison, M.H. (1968): Serotonin, norepinephrine and 3,4-dihydroxyphenylethylamine in rat brain parts following electroconvulsive shock. *Journal of Psychiatric Research, 6:* 143-152.

Hirsch, S.R., Gaind, R., Rohde, P.D., Stevens, B.C., and Wing, J.K. (1973): Outpatient maintenance of chronic schizophrenic patients with long-acting fluphenazine: double-blind placebo trial. *British Medical Journal, 192:* 633-637.

Ho, A.K.S., Loh, H.H., Craves, F., Hitzeman, R.J., and Gershon, S. (1970): The effect of prolonged lithium treatment on the synthesis rat and turn-over of monoamines in brain regions of rats. *European Journal of Pharmacology, 10:* 72-78.

Hökfelt, T., Fuxe, K., Johansson, O., and Ljungdahl, Å. (1974): Pharmacohistochemical evidence of the existence of dopamine nerve terminal in the rat limbic cortex. *European Journal of Pharmacology, 25:* 108-112.

148   AFFECTIVE DISORDERS

Hollister, L.E., Berger, P., Ogie, F.L., Arnold, R.C., and Johnson, A. (1974): Protirelin (TRH) in depression. *Archives of General Psychiatry, 31:* 468-470.

Horst, W.D., and Spirt, N. (1974): A possible mechanism for the antidepressant activity of thyrotropin releasing hormone. *Life Sciences, 15:* 1073-1082.

Hullin, R.P., Bailey, A.D., McDonald, R., Dransfield, G.A., and Milne, H.B. (1967): Variations in 11-hydroxycorticosteroids in depression and manic-depressive psychosis. *British Journal of Psychiatry, 113:* 593-600.

Hunter, K.R., Boakes, A.J., Laurence, D.R., and Stern, G.M. (1970): Monoamine oxidase inhibitors and L-dopa. *British Medical Journal, 3:* 388.

Iversen, L.L. (1973): Catecholamines. *British Medical Bulletin, 29.* Medical Department. London.

Janowski, D.S., El-Yousef, M.K., Davis, J.M., and Sekerke, H.J. (1972): A cholinergic adrenergic hypothesis of mania and depression. *Lancet, II:* 632-635.

Janowski, D.S., El-Yousef, M.K., Davis, J.M., and Sekerke, H.J. (1973): Parasympathetic suppression of manic symptoms by physostigmine. *Archives of General Psychiatry, 28:* 542-552.

Janowski, D.S., El-Yousef, M.K., and Davis, J.M. (1974): Acetylcholine and depression. *Psychosomatic Medicine, 36:* 248-257.

Jensen, K., Fruensgaard, K., Ahlfors, U.G., Pihkanen, T.A., Tuomikoski, S., Ose, E., Dencker, S.J., Lindberg, D., and Nagy, A. (1975): Tryptophan/imipramine in depression. *Lancet, II,* 920.

Jepson, J.B., Lovenberg, W., and Zaltzman, P. (1960): Amine metabolism studied in normal and phenylketonuric humans by monoamine oxidase inhibition. *Biochemical Journal, 74:* 5.

Jéquier, E., Lovenberg, W., and Sjoerdsma, A. (1967): Tryptophan hydroxylase inhibition: the mechanism by which p-chlorophenylalanine depletes rat brain serotonin. *Molecular Pharmacology, 3:* 274-278.

Jimerson, D.C., Gordon, E.K., Post, R.M., and Goodwin, F.K. (1975): Central noradrenergic function in man: vanillylmandelic acid in CSF. *Brain Research, 99,* 434-439.

Johnson, P., Kitchin, A.H., Lowther, C.P., and Turner, R.W.D. (1966): Treatment of hypertension with methyldopa. *British Medical Journal, 1:* 133-137.

Jonsson, L.E., Angard, E., and Gunne, L.M. (1971): Blockade of intravenous amphetamine euphoria in man. *Clinical Pharmacology and Therapeutics, 12:* 889-896.

Jori, A., and Garattini, S. (1965): Interaction between imipramine-like agents and catecholamine-induced hyperthermia. *Journal of Pharmacy and Pharmacology, 17:* 480-488.

Kane, F.J. (1970): Treatment of mania with cinanserin, an anti-serotonin agent. *American Journal of Psychiatry, 126:* 1020-1023.

Kansal, P.C., Buse, J., Talbert, O.R., and Buse, M.G. (1972): Effect of l-dopa on plasma growth hormone, insulin, and thyroxine. *Journal of Clinical Endocrinology and Metabolism, 34:* 99-105.

Karki, N.T. (1956): Urinary excretion of noradrenaline and adrenaline in different age groups. Its diurnal variation and effect of muscular work on it. *Acta Physiologica Scandinavica, 39 (suppl. 132):* 5-96.

Kastin, A.J., Ehrensing, R.H., Schalch, D.S., and Anderson, M.S. (1972): Improvement in mental depression with decreased thyrotropin response after administration of thyrotropin-releasing hormone. *Lancet, II:* 740-742.

Katz, R.I., Chase, T.N., and Kopin. I.J. (1968): Evoked release of norepinephrine and serotonin from brain slices: inhibition by lithium. *Science, 162:* 466-467.

Keller, H.H., Bartholini, G., and Pletscher, A. (1973): Increase of 3-methoxy-4-hydroxphenylethylene glycol in rat brain by neuroleptic drugs. *European Journal of Pharmacology, 23:* 183-186.

Keller, H.H., Bartholini, G., and Pletscher, A. (1974a): Enhancement of cerebral noradrenaline turnover by thyrotropin-releasing hormone. *Nature, 248:* 528-529.

Keller, H.H., Bartholini, G. and Pletscher, A. (1974b): Enhancement of noradrenaline turnover in rat brain by L-dopa. *Journal of Pharmacy and Pharmacology, 26:* 649-651.

Kety, S.S., Javoy, F., Thierry, A.M., Julou, L., and Glowinski, J. (1967): A sustained effect of electroconvulsive shock on the turnover of norepinephrine in the central nervous system of the rat. *Proceedings of the National Academy of Sciences, 58:* 1249-1254.

Kety, S.S (1971): Brain amines and affective disorders. In: *Brain Biochemistry and Mental Disease.* Edited by B.T. Ho and W.M. McIsaac, Plenum Press, New York. Pp. 237-263.

Klaiber, E.L., Kobayashi, Y., Broverman, D.M., and Hall, F. (1971): Plasma monoamine oxidase activity in regularly menstruating women and in amenorrheic women receiving cyclic treatment with estrogenes and a progestin. *Journal of Clinical Endocrinology, 33:* 630-638.

Klaiber, E.L., Broverman, D.M., Vogel, W., Kobayashi, Y., and Moriarty, D. (1972): Effects of estrogen therapy on plasma MAO activity and EEG driving responses of depressed women. *American Journal of Psychiatry, 128:* 42-48.

Kleinberg, D.L., Noel, G.L., and Frantz, A.G. (1971): Chlorpromazine stimulation and l-dopa suppression of plasma prolactin in man. *Journal of Clinical Endocrinology and Metabolism, 33:* 873-876.

Klerman, G.L., Schildkraut, J.J., Hasenbush, L.L., Greenblatt, M., and Friend, D.G. (1963): Clinical experience with dihydroxyphenylalanine (dopa) in depression. *Journal of Psychiatric Research, 1:* 289-297.

Kline, N.S., Sacks, W., and Simpson, G.M. (1964): Further studies on one day treatment of depression with 5-HTP. *American Journal of Psychiatry, 121:* 379-381.

Knox, W.E. (1951): Two mechanisms which increase in vivo the liver tryptophan peroxidase activity: specific enzyme adaptation and stimulation of the pituitary adrenal system. *British Journal of Experimental Pathology, 32:* 462-469.

Knox, W.E., and Auerbach, V.H. (1955): The hormonal control of tryptophan peroxidase in the rat. *Journal of Biological Chemistry, 214:* 307-313.

Koe, B.K., and Weissman, A. (1966a): Marked depletion of brain serotonin by p-chlorophenylalanine. *Federation Proceedings, 25:* 452.

Koe, B.K., and Weissman, A. (1966b): p-Chlorophenylalanine: specific depletor of brain serotonin. *Journal of Pharmacology and Experimental Therapeutics, 154:* 499-516.

Kopin, I.J., and Axelrod, J. (1963): The role of monoamine oxidase in the release and metabolism of norepinephrine. *Annals of the New York Academy of Sciences, 107:* 848-853.

Kopin, I.J. (1964): Storage and metabolism of catecholamines: the role of monoamine oxidase. *Pharmacological Reviews, 16:* 179-191.

Korf, J., and Praag, H.M. van. (1970): The intravenous probenecid test: a possible aid in evaluation of the serotonin hypothesis on the pathogenesis of depression. *Psychopharmacologia, 18:* 129-132.

Korf, J., Praag, H.M. van., and Sebens, J.B. (1971): Effect of intravenously administered probenecid in humans on the levels of 5-hydroxy-indole-acetic acid, homovanillic acid and 3-methoxy-4-hydroxy-phenyl-glycol in cerebrospinal fluid. *Biochemical Pharmacology, 20:* 659-668.

Korf, J., and Praag, H.M. van. (1971): Amine metabolism in human brain: further evaluation of the probenecid test. *Brain Research, 35:* 221-230.

Korf, J., and Praag, H.M. van. (1972): Action of p-chloroamphetamine on cerebral serotonin metabolism: an hypothesis. *Neuropharmacologia, 11:* 141-144.

Korf, J. Praag, H.M. van, and Sebens, J.B. (1972): Serum tryptophan decreased, brain tryptophan increased and brain serotonin synthesis unchanged after probenecid loading.

150   AFFECTIVE DISORDERS

Brain Research, 42: 239-242.

Korf, J., Schutte, H.H., and Venema, K. (1973): A semi-automated fluorometric determination of 5-hydroxyindoles in the nanogram range. *Analytical Biochemistry, 53:* 146-153.

Korf, J., Praag, H.M. van, Schut, T., Nienhuis, R.J., and Lakke, J.P.W.F. (1974a): Parkinson's disease: L-dopa therapy failure and amine metabolites in cerebrospinal fluid. *European Neurology, 12:* 340-350.

Korf, J., Venema, K., and Postema, F. (1974b): Decarboxylation of exogenous l-5-hydroxytryptophan after destruction of the cerebral raphe system. *Journal of Neurochemistry, 23:* 249-252.

Kuhar, M.J., Roth, R.H., and Aghajanian, G.K. (1972): Synaptosomes from forebrains of rats with midbrain raphe lesions: selective reduction of serotonin uptake. *Journal of Pharmacology and Experimental Therapeutics, 181:* 36-45.

Kupfer, D.J., and Bowers, M.B., Jr. (1972): REM sleep and central monoamine oxidase inhibition. *Psychopharmacologia, 27:* 183-190.

Kuriyama, K., and Speken, R. (1970): Effect of lithium on content and uptake of norepinephrine and 5-hydroxytryptamine in mouse brain synaptosomes and mitochondria. *Life Sciences, 9:* 1213-1220.

Lackroy, G.H., and Praag, H.M. van (1971): Lithium salts as sedatives. An investigation into the possible effect of lithium on acute anxiety. *Acta Psychiatrica Scandinavica, 47:* 163-173.

Ladisich, W., Steinhauff, N., and Matussek, N. (1969): Chronic administration of electroconvulsive shock and norepinephrine metabolism in the rat brain. *Psychopharmacologia, 15:* 296-304.

Lakke, J.P.W.F., Korf, J., Praag, H.M. van, and Schut, T. (1972): The predictive value of the probenecid test for the effect of levodopa therapy in Parkinson's disease. *Nature/New Biology, 236:* 208-209.

Lamprecht, F., Ebert, M.H., Turek, I., and Kopin, I.J. (1974): Serum dopamine-beta-hydroxylase in depressed patients and the effect of electroconvulsive shock treatment. *Psychopharmacologia, 40:* 241-248.

Lapin, I.P., and Oxenkrug, G.F. (1969): Intensification of the central serotonergic process as a possible determinant of the thymoleptic effect. *Lancet, I:* 132-136.

Levitt, M., Spector, S., Sjoerdsma, A., and Udenfriend, S. (1965): Elucidation of the rate-limiting step in norepinephrine biosynthesis in the perfused guinea pig heart. *Journal of Pharmacology and Experimental Therapeutics, 148:* 1-8.

Lewander, T., and Sjöström, R. (1973): Increase in the plasma concentration of free tryptophan caused by probenecid in humans. *Psychopharmacologia* (Berl), *33:* 81-86.

Lindvall, O., Björklund, A., Moore, R.Y., and Stenevi, U. (1974): Mesencephalic dopamine neurons projecting to neocortex. *Brain Research, 81:* 325-331.

Lingjaerde, O. (1963): Tetrabenazine (Nitoman) in the treatment of psychoses. *Acta Psychiatrica Scandinavica, 39 (suppl. 170):* 1.

Lipsett, D., Madras, B.K., Wurtman, R.J., and Munro, H.N. (1973): Serum tryptophan level after carbohydrate ingestion: selective decline in non-albumin-bound tryptophan coincident with reduction in serum free fatty acids. *Life Sciences, 12:* 57-64.

Lloyd, K.J., Farley, I.J., Deck, J.H.N., and Hornykiewicz, O. (1974): Serotonin and 5-hydroxyindoleacetic acid in discrete areas of the brainstem of suicide victims and control patients. *Advances in Biochemical Psychopharmacology, 11:* 387-397.

Loew, D., and Taeschler, M. (1966): Die Wirkung tricyclischer Antidepressiva auf die 5-hydroxytryptophan-Hyperthermie des Kaninchens. *Naunyn-Schmiedeberg's Archives of Experimental Pathology, 252:* 399-406.

Lopez-Ibor, A.J.J., Gutierrez, J.L.A., and Iglesias, M.L.M. (1973): Tryptophan and amitryptiline in the treatment of depression. A double blind study. *International Pharmacopsychiatry, 8:* 145-151.

Lorens, S.A., and Guldberg, H.C. (1974): Regional 5-hydroxytryptamine following selective midbrain raphe lesions in the rat. *Brain Research, 78:* 45-56.

Loveless, A.H., and Maxwell, D.R. (1965): A comparison of the effects of imipramine, trimipramine and some other drugs in rabbits treated with a monoamine oxidase inhibitor. *British Journal of Pharmacology, 25:* 158-170.

Lovenberg, W., Jéquier, E., and Sjoerdsma, A. (1968): Tryptophan hydroxylation in mammalian systems. *Advances in Pharmacology, 6A:* 21-36.

Lundborg, P. (1963): Storage function and amine levels of the adrenal medullary granules at various intervals after reserpine treatment. *Experientia, 19:* 479-480.

Maas, J.W., Fawcett, J.A., and Dekirmenjian, H. (1968): 3-Methoxy-4-hydroxyphenylglycol (MHPG) excretion in depressive states: pilot study. *Archives of General Psychiarty, 19:* 129-134.

Maas, J.W., Dekirmenjian, H., and Fawcett, J. (1971): Catecholamine metabolism, depression and stress. *Nature, 230:* 330-331.

Maas, J.W., Fawcett, J.A., and Dekirmenjian, H. (1972): Catecholamine metabolism, depressive illness, and drug response. *Archives of General Psychiatry, 26:* 252-262.

Maas, J.W., Dekirmenjian, H., and Jones, F. (1973a): The identification of depressed patients who have a disorder of ne metabolism and/or disposition. In: *Frontiers in Catecholamine Research.* Edited by E. Usdin and S. Snyder. Pergamon Press, New York. Pp. 1091-1096.

Maas, J.W., Dekirmenjian, H., Garver, D., Redmon, D.E., Jr., and Landis, D.H. (1973b): Excretion of catecholamine metabolites following intraventricular injection of 6-hydroxydopamine in the macaca speciosa. *European Journal of Pharmacology, 23:* 121-130.

McCabe, M.S., Reich, T., and Winokur, G. (1970): Methysergide as a treatment for mania. *American Journal of Psychiatry, 127:* 354-356.

Maclean, R., Nicholson, W.J.N., Pare, C.M.B., and Stacey, R.S. (1965): Effect of monoamineoxidase inhibitors on the concentrations of 5-hydroxytryptamine in the human brain. *Lancet, II:* 205-208.

McNamee, H.B., Moody, J.P., and Naylor, G.J. (1972a): Indoleamine metabolism in affective disorders: Excretion of tryptamine indoleacetic acid and 5-hydroxyindoleacetic acid in depressive states. *Journal of Psychosomatic Research, 16:* 63-70.

McNamee, H.B., Le Poidevin, D., and Naylor, G.J. (1972b): Methysergide in mania: a double-blind comparison with thioridazine. *Psychological Medicine, 2:* 66-69.

Manara, L., Algeri, S., and Sestini, M.G. (1967): Some modifications of the adrenergic mechanism induced by DMI-reserpine interactions. In: *Antidepressant Drugs.* Proceedings of the First International Symposium. Edited by S. Garattini and M.N.G. Dukes. Excerpta Medica Foundation, Amsterdam. Pp. 51-60.

Mandell, A.J. (1975): Neurobiological mechanisms of presynaptic metabolic adaptation and their organization: implications for a pathophysiology of the affective disorders. In: *Neurobiological Mechanism of Adaptation and Behavior.* Edited by A.J. Mandell. Raven Press, New York.

Mangoni, A. (1974): The "kynurenine shunt" and depression. *Advances in Biochemical Psychopharmacology, 11:* 293-298.

Martin, J.B. (1973): Neural regulation of growth hormone secretion. *New England Journal of Medicine, 288:* 1384-1393.

Masuoka, D.T., Schott, H.F., and Petriello, L. (1963): Formation of catecholamines by various areas of cat brain. *Journal of Pharmacology and Experimental Therapeutics, 139:* 73-76.

Matussek, N., Pohlmeier, H., and Rüther, E. (1966): Die Wirkung von Dopa auf gehemmte Depressionen. *Klinische Wochenschrift, 44:* 727-728.

Matussek, N., Benkert, O., Schneider, K., Otten, H., and Pohlmeier, H. (1970): Wirkung eines Decarboxylasehemmers (Ro 4-4602) in Kombination mit L-dopa auf gehemmte Depressionen. *Arzneimittelforschung, 20:* 934-937.

Meek, J.L., and Neff, N.H. (1973): Is cerebrospinal fluid the major avenue for the removal of 5-hydroxyindoleacetic acid from the brain? *Neuropharmacology, 12:* 497-499.

Meites, J., Lu, K.H., Wuttke, W., Welsch, C.W., Nagasawa, H., and Quadrie, F.K (1972): Recent studies on function and control of prolactin secretion in rats. *Recent Progress in Hormone Research, 28:* 471-526.

Mendels, J. (1971): Relationship between depression and mania. *Lancet, I:* 342.

Mendels, J., Frazer, A., Fitzgerald, R.G., Ramsey, T.A., and Stokes, J.W. (1972): Biogenic amine metabolites in cerebrospinal fluid of depressed and manic patients. *Science, 175:* 1380-1382.

Mendels, J., and Frazer, A.J. (1973): Intracellular lithium concentration and clinical response: towards a membrane theory of depression. *Journal of Psychiatric Research, 10:* 9-18.

Mendels, J., Stinnet, J.L., Burns, D., and Frazer, A. (1975): Amine precursors and depression. *Archives of General Psychiatry, 32:* 22-30.

Messiha, F.S., Agallianos, D., and Clower, C. (1970): Dopamine excretion in affective states and following Li_2Co_3 therapy, *Nature, 225:* 868-869.

Miller, E.E., Sawano, S., and Arimura, A. (1967): Blockade of release of growth hormone by brain norepinephrine depletors. *Endocrinology, 80:* 471-476.

Modigh, K. (1972): Central and peripheral effects of 5-hydroxytryptophan on motor activity in mice. *Psychopharmacologia, 23:* 48-54.

Modigh, K. (1973a): Effects of l-tryptophan on motor activity in mice. *Psychopharmacologia, 30:* 123-134.

Modigh, K. (1973b): Effects of chloroimipramine and protriptyline on the hyperactivity induced by 5-hydroxytryptophan after peripheral decarboxylase inhibition in mice. *Journal of Neural Transmission, 34:* 101-109.

Moir, A.T.B., Ashcroft, G.W., Crawford, T.B.B., Eccleston, D., and Guildberg, H.C. (1970): Cerebral metabolites in cerebrospinal fluid as a biochemical approach to the brain. *Brain, 93:* 357-368.

Morgan, W.W., Saldana, J.J., Yndo, C.A., and Morgan, J.F. (1975): Correlations between circadian changes in serum amino acids or brain tryptophan and the contents of serotonin and 5-hydroxyindoleacetic acid in regions of the rat brain. *Brain Research, 84:* 75-86.

Mosnaim, A.D., Inwang, E.E., Sugerman, J.H. DeMartini, W.J., and Sabelli, H.C. (1973): Ultraviolet spectrophotometric determination of 2-phenylethylamine in biological samples and its possible correlation with depression. *Biological Psychiatry, 6:* 235-257.

Mosnaim, A.D., Inwang, E.E., and Sabelli, H.C. (1974): The influence of psychotropic drugs on the levels of endogenous 2-phenylethylamine in rabbit brain. *Biological Psychiatry, 8:* 227-234.

Montjoy, C.Q., Price, J.S., Weller, M., Hunter, P., Hall, R., and Dewar, J.H. (1974): A double-blind cross-over sequential trial of oral thyrotropin-releasing hormone in depression. *Lancet, I:* 958-960.

Müller, E.E., Pecile, A., Felici, M., and Cocchi, D. (1970): Norepinephrine and dopamine injection into lateral brain ventricles of the rat and growth hormone releasing activity in hypothalamys and plasma. *Endocrinology, 86:* 1376-1382.

Murphy, D.L., Colburn, R.W., Davis, J.M., and Bunney, W.E., Jr. (1969): Stimulation by lithium of monoamine uptake in human platelets. *Life Sciences, 8:* 1187-1193.

Murphy, D.L., Colburn, R.W., Davis, J.M., and Bunney, W.E., Jr. (1970): Imipramine and lithium effects on biogenic amine transport in depressed and manic-depressive patients. *American Journal of Psychiatry, 127:* 339-345.

Murphy, D.L., Brodie, H.K.H., Goodwin, F.K., and Bunney, W.E., Jr. (1971): L-dopa: regular induction of hypomania in "bipolar" manic depressive patients. *Nature, 229:* 135-136.

Murphy, D.L., and Weiss, R. (1972): Reduced monoamine oxidase activity in blood platelets from bipolar depressed patients. *American Journal of Psychiatry, 128:* 35-41.

Murphy, D.L. (1972): Amine precursors, amines, and false neurotransmitters in depressed patients. *American Journal of Psychiatry, 129:* 141-148.

Murphy, D.L., Baker, M., Goodwin, F.K., Miller, H., Kotin, J., and Bunney, W.E., Jr. (1974): L-tryptophan in affective disorders: indoleamine changes and differential clinical effects. *Psychopharmacologia, 34:* 11-20.

Musacchio, J.M., Julou, L., Kety, S.S., and Glowinski, J. (1969): Increase in rat brain tyrosine hydroxylase activity produced by electroconvulsive shock. *Proceedings of the National Academy of Sciences of the United States of America, 63:* 1117-1119.

Nagatsu, T., Hidaka, H., Kuzuya, H., and Takeya, K. (1970): Inhibition of dopamine-β-hydroxylase by fusaric acid (5-butylpicolinic acid) in vitro and in vivo. *Biochemical Pharmacology, 19:* 35-44.

Neff, N.H., Tozer, T.N., and Brodie, B.B. (1967): Application of steady-state kinetics to studies of the transfer of 5-hydroxyindoleacetic acid from brain to plasma. *Journal of Pharmacology and Experimental Therapeutics, 158:* 214-218.

Neff, N.H., and Goridis, C. (1972): Neuronal monoamine oxidase specific enzyme types and their rates of formation. *Advances in Biochemical Psychopharmacology, 5:* 307-323.

Neff, N.H., and Yang, H.Y.T. (1974): Another look at the monoamine oxidases and the monoamine oxidase inhibitor drugs. *Life Sciences, 14:* 2061-2074.

Ng, K.Y., Chase, T.N., Colburn, R.W., and Kopin, I.J. (1970): L-dopa induced release of cerebral monoamines. *Science, 170:* 76-77.

Nordin, G., Ottosson, J.O., and Roos, B-E. (1971): Influence of convulsive therapy on 5-hydroxyindoleacetic acid and homovanillic acid in cerebrospinal fluid in endogenous depression. *Psychopharmacologia, 20:* 315-320.

Nybäck, H., and Sedvall, G. (1970): Further studies on the accumulation and disappearance of catecholamines formed from tyrosine-^{14}C in mouse brain. Effect of some phenothiazine analogues. *European Journal of Pharmacology, 10:* 193-205.

Olsson, R., and Roos, B-E. (1968): Concentrations of 5-hydroxyindoleacetic acid and homovanillic acid in the cerebrospinal fluid after treatment with probenecid in patients with Parkinson's disease. *Nature, 219:* 502-503.

Papeschi, R., and McClure, D.J. (1971): Homovanillic acid and 5-hydroxyindoleacetic acid in cerebrospinal fluid of depressed patients. *Archives of general Psychiatry, 25:* 354-358.

Papeschi, R., Sourkes, T.L., Poirier, L.J., and Boucher, R. (1971): On the intracerebral origin of homovanillic acid of the cerebrospinal fluid of experimental animals. *Brain Research, 28:* 527-533.

Papeschi, R., Randrup, A., and Munkvad, I. (1974): Effect of ECT on dopaminergic and noradrenergic mechanisms. *Psychopharmacologia, 35:* 159-168.

Pare, C.M.B., and Sandler, M. (1959): A clinical and biochemical study of a trial of iproniazid in the treatment of depression. *Journal of Neurology, Neurosurgery and Psychiatry, 22:* 247-251.

Pare, C.M.B. (1963): Potentiation of monoamine oxidase inhibitors by tryptophan. *Lancet, II:* 527-528.

Pare, C.M.B., Yeung, D.P.H., Price, K., and Stacey, R.S. (1969): 5-Hydroxytryptamine in brainstem, hypothalamus and caudate nucleus of controls and of patients committing suicide by coal-gas poisoning. *Lancet, II:* 133-135.

Paul, M.J., Ditzion, B.R., Pauk, G.L., and Janowski, D.S. (1970): Urinary adenosine 3'5'-monophosphate excretion in affective disorders. *American Journal of Psychiatry, 126:* 1493-1497.

Paul, M.J., Cramer, H., and Goodwin, F.K. (1971): Urinary cyclic AMP excretion in depression and mania. *Archives of General Psychiatry, 24:* 327-333.

Paykel, E.S. (1972): Correlates of a depressive typology, *Archives of General Psychiatry, 27:* 203-210.

Perel, J.M., Black, N., Wharton, R.N., and Malitz, S. (1969): Inhibition of imipramine metabolism by methylphenidate. *Federation Proceedings, 28:* 418.

Perez-Cruet, J., Tagliamonte, P., Tagliamonte, A., and Gessa, G.L. (1971): Stimulation of serotonin synthesis by lithium. *Journal of Pharmacology and Experimental Therapeutics, 178:* 325-330.

Perris, C. (1966): A study of bipolar (manic-depressive) and unipolar recurrent depressive psychoses. *Acta Psychiatrica Scandinavica, 42 (suppl. 194):* 1-189.

Persson, T., and Roos, B-E. (1968): 5-Hydroxytryptophan for depression. *Lancet, II:* 987-988.

Pflanz, G., and Palm, D. (1973): Acute enhancement of dopamine-β-hydroxylase activity in human plasma after maximum work load. *European Journal of Clinical Pharmacology, 5:* 555-558.

Pletscher, A., Gey, K.F., and Zeller, P. (1960): Monoamine-Oxidase-Hemmer. *Progress in Drug Research, 2:* 417-590.

Pletscher, A., Bartholini, G., Bruderer, H., Burkard, W.P., and Gey, K.F. (1964): Chlorinated aralkylamines affecting the cerebral metabolism of 5-hydroxytryptamine. *Journal of Pharmacology and Experimental Therapeutics, 145:* 344-350.

Pletscher, A., Gey, K.F., and Burkard, W.P. (1966a): Inhibitors of monoamine oxidase and decarboxylase of aromatic acids. *Handbook of Experimental Pharmacology, 19:* 593-735.

Pletscher, A., da Prada, M., Burkard, W.P., Bartholini, G., Steiner, F.A., Bruderer, H., and Bigler, F. (1966b): Aralkylamines with different effects on the metabolism of aromatic monoamines. *Journal of Pharmacology and Experimental Therapeutics, 154:* 64-72.

Pletscher, A., Bartholini, G., and Tissot, R. (1967): Metabolic fate of 1-(^{14}C) dopa in cerebrospinal fluid and blood plasma of humans. *Brain Research, 4:* 106-109.

Pletscher, A. (1968a): Monoamine oxidase inhibitors: effects related to psychostimulation. In: *Psychopharmacology. A Review of Progress.* Edited by D.H. Efron. Public Health Service, Washington. Pp. 649-654.

Pletscher, A. (1968b): Metabolism, transfer and storage of 5-HT (5-hydroxytryptamine) in blood platelets. *British Journal of Pharmacology and Chemotherapy, 32:* 1-16.

Plotnikoff, N.P., Prange, A.J., Jr., Breese, G.R., Anderson, M.S., and Wilson, I.C. (1972): Thyrotropin releasing hormone: enhancement of dopa activity by a hypothalamic hormone. *Science, 178:* 417-481.

Plotnikoff, N.P., Prange, A.J., Jr., Breese, G.R., and Wilson, I.C. (1974): Thyrotropin releasing hormone: enhancement of dopa activity in thyroidectomized rats. *Life Sciences, 14:* 1271-1278.

Poitou, P., Guerinot, F., and Bohuon, C. (1974): Effect of lithium on central metabolism of 5-hydroxytryptamine. *Psychopharmacologia, 38:* 75-80.

Pöldinger, W. (1963): Combined administration of desipramine and reserpine or tetrabenazine in depressed patients. *Psychopharmacologia, 4:* 308.

Post, R.M., Goodwin, F.K., Gordon, E. and Watkin, D.M. (1973a): Amine metabolites in human cerebrospinal fluid: Effects of cord transaction and spinal fluid block. *Science, 179:* 897-899.

Post, R.M., Kotin, J., Goodwin, F.K., and Gordon, E.K. (1973b): Psychomotor activity and cerebrospinal fluid amine metabolites in affective illness. *American Journal of Psychiatry, 130:* 67-72.

Post, R.M., Gordon, E.K, Goodwin, F.K., Bunney, W.E., Jr. (1973c): Central norepinephrine metabolism in affective illness: MHPG in the cerebrospinal fluid. *Science, 179:* 1002-1003.

Post, R.M., and Goodwin, F.K. (1973): Stimulated behavior states: an approach to specificity in psychobiological research. *Biological Psychiatry, 7:* 237-254.

Post, R.M., and Goodwin, F.K. (1974): Effects of amitryptiline and imipramine on amine metabolites in the cerebrospinal fluid of depressed patients. *Archives of General Psychiatry, 30:* 234-239.

Potter, W.P. de, Schaepdryver, A.F. de, Moerman, E.J. and Smith, A.D. (1969): Evidence for the release of vesicle proteins together with noradrenaline upon stimulation of the splenic nerve. *Journal of Physiology, 204:* 102p-104p.

Praag, H.M. van (1962): A critical investigation of the importance of MAO inhibition as a therapeutic principle in the treatment of depression. Thesis, Utrecht.

Praag, H.M. van, and Leijnse, B. (1963a): Die Bedeutung der Monoaminoxydasehemmung als antidepressives Prinzip. I. *Psychopharmacologia, 4:* 1-14.

Praag, H.M. van, and Leijnse, B. (1963b): Die Bedeutung der Psychopharmacologie für die klinische Psychiatrie. Systematik als notwendiger Ausgangspukt. *Nervenarzt, 34:* 530-537.

Praag, H.M. van, and Leijnse, B. (1965): Neubewertung des Syndrome. Skizze einer funktionellen Pathologie. *Psychiatria, Neurologia, Neurochirurgia, 68:* 50-66.

Praag, H.M. van, Uleman, A.M., and Spitz, J.C. (1965): The vital syndrome interview. A structured standard interview for the recognition and registration of the vital depressive symptom complex. *Psychiatria, Neurologia, Neurochirurgia, 68:* 329-346.

Praag, H.M. van (1967): Antidepressants, catecholamines and 5-hydroxyindoles. Trends towards a more specific research in the field of antidepressants. *Psychiatria, Neurologia, Neurochirurgia, 70:* 219-233.

Praag, H.M. van (1969): Depressie en de stofwisseling van 5-hydroxytryptamine. *Nederlands Tijdschrift voor Geneeskunde, 113:* 2245-2247.

Praag, H.M. van, Korf, J., and Puite, J. (1970): 5-Hydroxyindoleacetic acid levels in the cerebrospinal fluid of depressive patients treated with probenecid. *Nature, 225:* 1259-1260.

Praag, H.M. van, and Korf, J. (1970): L-tryptophan in depression. *Lancet, II:* 612.

Praag, H.M. van, and Korf, J. (1971a): Endogenous depressions with and without disturbances in the 5-hydroxytryptamine metabolism: a biochemical classification? *Psychopharmacologia, 19:* 148-152.

Praag, H.M. van, and Korf, J. (1971b): Nieuwe ontwikkelingen op het terrein van de antidepressiva. *Nederlands Tijdschrift voor Geneeskunde, 115:* 1963-1970.

Praag, H.M. van, and Korf, J. (1971c): Retarded depressions and the dopamine metabolism. *Psychopharmacologia, 19:* 199-203.

Praag, H.M. van, Korf, J., Dols, L.C.W., and Schut, T. (1972): A pilot study of the predictive value of the probenecid test in application of 5-hydroxytryptophan as an antidepressant. *Psychopharmacologia, 25:* 14-21.

Praag, H.M. van (1972): Biological psychiatry in perspective. The dangers of sectarianism in psychiatry. *Comprehensive Psychiatry, 13:* 401-410.

Praag, H.M. van, Korf, J., and Schut, T. (1973a): Cerebral monoamines and depression. An investigation with the probenecid technique. *Archives of General Psychiatry, 28:* 827-831.

Praag, H.M. van, and Korf, J. (1973): 4-Chloramphetamines. Chance and trend in the development of new antidepressants. *Journal of Clinical Pharmacology, 13:* 3-14.

Praag, H.M. van, Flentge, F., Korf, J., Dols, L.C.W. and Schut, T. (1973b): The influence of probenecid in the metabolism of serotonin, dopamine and their precursors in man. *Psychopharmacologia, 33:* 141-151.

Praag, H.M. van (1974a): Towards a biochemical typology of depressions? *Pharmacopsychiatry, 7:* 281-292.

Praag, H.M. van (1974b): New developments in human psychopharmacology. *Comprehensive Psychiatry, 15:* 389-401.

Praag, H.M. van, Burg, W. van den, Bos, E.R.H. and Dols, L.C.W. (1974): 5-Hydroxy-tryptophan in combination with clomipramine in "therapy-resistant" depression. *Psychopharmacologia, 38:* 267-269.

Praag, H.M. van, and Korf, J. (1974): 5-Hydroxytryptophan as an antidepressant. *Journal of Nervous and Mental Disease, 158:* 331-337.

Praag, H.M. van (1975): Neuroleptics as a guideline to biological research in psychotic disorders. *Comprehensive Psychiatry, 16:* 7-22.

Praag, H.M. van, and Korf, J. (1975a): 4-Chloramphetamines. In: *Psychotherapeutic Drugs.* Edited by E. Usdin and I.S. Forrest. Dekker, New York.

Praag, H.M. van, Dols, L.C.W., and Schut, T. (1975a): Biochemical versus psychopathological action profile of neuroleptics. A comparative study of chlorpromazine and oxyper-tine in acute psychotic disorders. *Comprehensive Psychiatry, 16:* 255-263.

Praag, H.M. van, and Korf, J. (1975b): Biochemical research into psychosis. Results of a new strategy. *Acta Psychiatrica Scandinavia, 51:* 268-284.

Praag, H.M. van, Korf, J., Lakke, J.W.F. and Schut, T. (1975b): Dopamine metabolism in depressions, psychoses, and Parkinson's disease: The problem of the specificity of biological variables in behaviour disorders. *Psychological Medicine, 5:* 138-146.

Praag, H.M. van, and Korf, J. (1975c): Neuroleptics, catecholamines and psychotic dis-orders. A study of their interrelation. *American Journal of Psychiatry, 132:* 593-597.

Praag, H.M. van, and Korf, J. (1975d): Central monoamine deficiency in depression: causative or secondary phenomenon? *Pharmakopsychiatrie, 8:* 321-326.

Praag, H.M. van (1976): The vulnerable brain. Biological factors in the diagnosis and treat-ment of depression. In: *Psychiatric diagnosis: new evidence for reconsideration of existing categories.* Brunner/Mazel, New York.

Prange, A., Pustrom, E , and Cochrane, C. (1964): Imipramine enhancement of norepine-phrine in normal humans. *Psychiatry Digest, 25:* 27-40.

Prange, A.J., Jr., McCurdy, L.R., and Cochrane, C.M. (1967): The systolic blood pressure response of depressed patients to infused norepinephrine. *Journal of Psychiatric Re-search, 5:* 1-13.

Prange, A.J., Jr., Wilson, I.C., Knox, A., McClane, T.K., and Lipton, M.A. (1970): Enhance-ment of imipramine by thyroid stimulating hormone: clinical and theoretical implica-tions. *American Journal of Psychiatry, 127:* 191-199.

Prange, A.J., Jr., Wilson, I.C., Lara, P.P., Alltop, L.B., and Breese, G.R. (1972a): Effects of thyrotropin-releasing hormone in depression. *Lancet, II:* 999-1002.

Prange, A.J., Jr., Wilson, I.C., Knox, A.E., McClane, T.K., Breese, G.R., Martin, B.R., Alltop, L.B., and Lipton, M.A. (1972b): Thyroid-imipramine clinical and chemical interaction: evidence for a receptor deficit in depression. *Journal of Psychiatric Re-search, 9:* 187-205.

Prange, A.J., Jr., Wilson, I.C., Lynn, C.W., Alltop, L.B., Strikeleather, R.A., and Raleigh, N.C. (1974): L-tryptophan in mania. *Archives of General Psychiatry, 30:* 52-62.

Prockop, L., Fahn, S., and Barbour, P. (1974): Homovanillic acid: entry rate kinetics for transfer from plasma to cerebrospinal fluid. *Brain Research, 80:* 435-442.

Prichard, B.N.C., Johnston, A.W., Hill, I.D., and Rosenheim, M.L. (1968): Bethanidine, guanethidine, and methyldopa in the treatment of hypertension: a within-patient comparison. *British Medical Journal, 1:* 135-144.

Puite, J.K., Schut, T., Praag, H.M. van, and Lakke, J.P.W.F. (1973): Monoamine metabolism and depression in Parkinson patients. *Psychiatria, Neurologia, Neurochirurgia, 76:* 61-70.

Ramsden, E.N. (1970): Cyclic AMP in depression and mania. *Lancet, II:* 108.

Redmond, E.E., Jr., Maas, J.W., Kling, A., Graham, C.W., and Dekirmenjian, H. (1971): Social behavior of monkeys selectively depleted of monoamines. *Science, 174:* 428-430.

Rees, J.R., Alltopp, M.N.E., and Hullin, R.P. (1974): Plasma concentrations of tryptophan and other amino acids in manic-depressive patients. *Psychological Medicine, 4:* 334-337.

Reigle, T.G., Avni, J., Platz, P.A., Schildkraut, J.J., and Plotnikoff, N.P. (1974): Norepinephrine metabolism in the rat brain following acute and chronic administration of thyrotropinreleasing hormone. *Psychopharmacologia, 37:* 1-6.

Rimón, R., and Räkköläinen, V. (1968): Lithium iodide in the treatment of confusional states. *British Journal of Psychiatry, 114:* 109-110.

Rimón, R., Roos, B-E., Räkköläinen, V., and Alanen, Y. (1971): The content of 5-hydroxyindoleacetic acid and homovanillic acid in the cerebrospinal fluid of patients with acute schizophrenia. *Journal of Psychosomatic Research, 15:* 375-378.

Robison, G.A., Coppen, A.J., Whybrow, P.C., and Prange, A.J., Jr. (1970): Cyclic AMP in affective disorders. *Lancet, II:* 1028-1029.

Robins, E., Munoz, R.A., Martin, S., and Gentry, K.A. (1972): In: *Disorders of Mood.* Edited by J. Zubin and F.A. Freyhan, John Hopkins Press, Baltimore. Pp. 33-45.

Robinson, D.S., Davis, J.M., Nies, A., Ravaris, C.G., Sylwester, D. (1971): Relation of sex and aging to monoamine oxidase activity of human brain, plasma and platelets. *Archives of General Psychiatry, 24:* 536-539.

Robinson, D.S., Davis, J.M., Nies, A., Colburn, R.W., Davis, J.N., Bourne, H.R., Bunney, W.E., Jr., Shaw, D.M., and Coppen, A.J. (1972): Ageing, monoamines, and monoamine oxidase levels. *Lancet, I:* 290-291.

Roos, B-E., and Sjöström, R. (1969): 5-Hydroxyindoleacetic acid and homovanillic acid levels in the cerebrospinal fluid after probenecid application in patients with manicdepressive psychosis. *Journal of Clinical Pharmacology, 1:* 153-155.

Rossloff, B.N., and Davis, J.M. (1974): Effect of iprindole on norepinephrine turnover and transport. *Psychopharmacologia, 40:* 53-64.

Ross, S.B., and Renyi, A.L. (1966): In vivo inhibition of H³- noradrenaline uptake by mouse brain slices in vitro. *Journal of Pharmacy and Pharmacology, 18:* 322-323.

Rossum, J.M. van (1967): The significance of dopamine receptor blockade for the action of neuroleptic drugs. In: *Neuropsychopharmacology.* Edited by H. Brill. Excerpta Medica Foundation, Den Haag. Pp. 321-329.

Roth, J.A., and Gillis, C.N. (1974a): Deamination of β-phenylethylamine by monoamine oxidase-inhibition by imipramine. *Biochemical Pharmacology, 23:* 2537-2545.

Roth, J.A., and Gillis, C.N. (1974b): Inhibition of lung, liver and brain monoamine oxidase by imipramine and desipramine. *Biochemical Pharmacology, 23:* 1138-1140.

Rubin, R.T. (1967): Adrenal cortical activity changes in manic-depressive illness. *Archives of General Psychiatry, 17:* 671-679.

Rubin, R.T., Miller, R.G., Clark, B.R., Poland, R.E., and Arthur, R.J. (1970): The stress of aircraft carrier landings II. 3-Methoxy-4-hydroxyphenylglycol excretion in naval aviators. *Psychosomatic Medicine, 32:* 589-597.

Rutledge, C.O. (1970): The mechanisms by which amphetamine inhibits oxidative deamination of norepinephrine in brain. *Journal of Pharmacology and Experimental Thera-*

peutics, 171: 188-195.

Sabelli, H.C., and Mosnaim, A.D. (1974): Phenylethylamine hypothesis of affective behavior. *American Journal of Psychiatry, 131:* 695-699.

Sack, R.L., and Goodwin, F.K. (1974): Inhibition of dopamine-β-hydroxylase in manic patients. *Archives of General Psychiatry, 31:* 649-654.

Sachar, E.J. (1967): Corticosteroids in depressive illness. I. A re-evaluation of control issues and the literature. *Psychosomatic Medicine, 30:* 162-171.

Sachar, E.J., Frantz, A.G., Altman, N., and Sassin, J. (1973): Growth hormone and prolactin in unipolar and bipolar depressed patients: responses to hypoglycemia and l-dopa. *American Journal of Psychiatry, 130:* 1362-1367.

Sanders-Bush, E., and Sulser, F. (1970): p-Chloroamphetamine: in vivo investigations on the mechanism of the action of the selective depletion of cerebral serotonin. *Journal of pharmacology and Experimental Therapeutics, 175:* 419-426.

Sanders-Bush, E., Bushing, J.A., and Sulser, F. (1972): p-Chloro-amphetamine-inhibition of cerebral tryptophan hydroxylase. *Biochemical Pharmacology, 21:* 1501-1510.

Sandler, M., and Youdim, M.B.H. (1972): Multiple forms of monoamine oxidase: functional, significance. *Pharmacology Reviews, 24:* 331-348.

Sandler, M., Bonham Carter, S., Cuthbert, M.F., and Pare, C.M.B. (1975): Is there an increase in monoamine-oxidase activity in depressive illness? *Lancet, I,* 1045-1048.

Sano, I. (1972). L-5-hydroxytryptophan (l-5-HTP)-therapie bei endogener Depression. *Münchener Medizinischer Wochenschrift, 144:* 1713-1716.

Schanberg, S.M., Schildkraut, J.J., and Kopin, I.J. (1967): Effects of psychoactive drugs on norepinephrine-H³ metabolism in brain. *Biochemical Pharmacology, 16:* 393-399.

Schanberg, S.M., Schildkraut, J.J., Breese, G.R., and Kopin, I.J. (1968): Metabolism of normetanephrine H³ in rat brain—identification of conjugated 3-methoxy-4-hydroxyphenylglycol as major metabolite. *Biochemical Pharmacology, 17:* 247-254.

Scheckel, C.L., and Boff, E. (1964): Behavioral effects of interacting imipramine and other drugs with d-amphetamine, cocaine and tetrabenazine. *Psychopharmacologia, 5:* 198-208.

Scheyen, J.D. van (1971): Behandeling van manie met methysergide. *Nederlands Tijdschrift voor Geneeskunde, 115:* 1634-1637.

Schildkraut, J.J., Klerman, G.L., Friend, D.G., and Greenblatt, M. (1963): Biochemical and pressor effects of oral D,l-dihydroxyphenylalanine in patients pretreated with antidepressant drugs. *Annals of the New York Academy of Sciences, 107:* 1005-1015.

Schildkraut, J.J., Klerman, G.L., Hammond, R., and Friend, D.G. (1964): Excretion of 3-methoxy-4-hydroxymandelic acid (VMA) in depressed patients treated with antidepressant drugs. *Journal of Psychiatric Research, 2:* 257-266.

Schildkraut, J.J. (1965): The catecholamine hypothesis of affective disorders: a review of supporting evidence. *American Journal of Psychiatry, 122:* 509-522.

Schildkraut, J.J., Green, R., Gordon, E.K., and Durell, J. (1966): Normetanephrine excretion and affective state in depressed patients treated with imipramine. *American Journal of Psychiatry, 123:* 690-700.

Schildkraut, J.J., Schanberg, S.M., Breese, G.R., and Kopin, I.J. (1967): Norepinephrine metabolism and drugs used in the affective disorders: a possible mechanism of action. *American Journal of Psychiatry, 124:* 600-608.

Schildkraut, J.J., and Kety, S.S. (1967): Biogenic amines and emotion. *Science, 156:* 21-30.

Schildkraut, J.J., Dodge, G.A., and Logue, M.A. (1969a): Effects of tricyclic antidepressants on uptake and metabolism of intracisternally administered norepinephrine-H³ in rat brain. *Journal of Psychiatric Research, 7:* 29-34.

Schildkraut, J.J., Logue, M.A., and Dodge, G.A. (1969b): Effects of lithium salts on turnover and metabolism of norepinephrine in rat brain. *Psychopharmacologia, 14:* 135-141.

Schildkraut, J.J., Draskoczy, P.R., and Sun Lo, P. (1971): Norepinephrine pools in rat brain: differences in turnover rates and pathways of metabolism. *Science, 172:* 587-589.

Schildkraut, J.J. (1973a): Pharmacology—the effects of lithium on biogenic amines. In: *Lithium. Its Role in Psychiatric Research and Treatment.* Edited by S. Gershon and B. Shopsin. Plenum Press, New York. Pp. 51-73.

Schildkraut, J.J. (1973b): Norepinephrine metabolites as biochemical criteria for classifying depressive disorders and predicting responses to treatment: preliminary findings. *American Journal of Psychiatry, 130:* 695-698.

Schildkraut, J.J., and Draskoczy, P.R. (1974): Effects of electroconvulsive shock on norepinephrine turnover and metabolism: basic and clinical studies. In: *Psychobiology of Convulsive Therapy.* Edited by M. Fink, S. Kety, J. McGaugh and T.A. Williams. J. Wiley and Sons, New York. Pp. 143-170.

Schildkraut, J.J. (1974): Biochemical criteria for classifying depressive disorders and predicting responses to pharmacotherapy; preliminary findings from studies of norepinephrine metabolism. *Pharmacopsychiatry, 7:* 98-107.

Schildkraut, J.J. (1975): Depressions and biogenic amines. In: *American Handbook of Psychiatry, VI.* Edited by D. Hamburg. Basic Books, New York.

Schou, M. (1973): Prophylactive lithium maintenance treatment in recurrent endogenous affective disorders. In: *Lithium. Its Role in Psychiatric Research and Treatment.* Edited by S. Gershon and B. Shopsin. Plenum Press, New York. Pp. 269-294.

Schubert, J., Nyback, H., and Sedvall, G. (1970): Effect of antidepressant drugs on accumulation and disappearance of monoamines formed in vivo from labelled precursors in mouse brain. *Journal of Pharmacy and Pharmacology, 22:* 136-139.

Schubert, J. (1973a): Metabolism of 5-hydroxytryptamine in brain and the effect of psychoactive drugs. Thesis, Stockholm.

Schubert, J. (1973b): Effect of chronic lithium treatment on monoamine metabolism in rat brain. *Psychopharmacologia, 32:* 301-311.

Schuckit, M., Robins, E., and Feighner, J. (1971): Tricyclic antidepressants and monoamine oxidase inhibitors. *Archives of General Psychiatry, 24:* 509-514.

Sebens, J.B., and Korf, J. (1975): Cyclic AMP in cerebrospinal fluid: accumulation following probenecid and biogenic amines. *Experimental Neurology, 46:* 333-344.

Segawa, T., and Kuruma, I. (1968): Influence of drugs on uptake of 5-hydroxytryptamine by nerve-ending particles of rabbit brain stem. *Journal of Pharmacy and Pharmacology, 20:* 320-322.

Sethna, E.R. (1974): A study of refractory cases of depressive illneses and their response to combined antidepressant treatment. *British Journal of Psychiatry, 124:* 265-272.

Shaw, D.M., Camps, F.E., and Eccleston, E.G. (1967): 5-Hydroxytryptamine in hind-brain of depressive suicides. *British Journal of Psychiatry, 113:* 1407-1411.

Shaw, D.M., O'Keeffe, R., Macsweeney, D.A., Brooksbank, B.W.L., Noguera, R., and Coppen, A. (1973): 3-Methoxy-4-hydroxyphenylglycol in depression. *Psychological Medicine, 3:* 333-336.

Shaw, D.M. (1975): Lithium and amine metabolism. In: *Lithium, Research and Therapy.* Edited by F.N. Johnson. Academic Press, New York. Pp. 411-423.

Sheard, M.H., and Aghajanian, G.K. (1968): Stimulation of the midbrain raphe: effect on serotonin metabolism. *Journal of Pharmacology and Experimental Therapeutics, 163:* 425-430.

Sheard, M., and Aghajanian, G.K. (1970): Neuronally activated metabolism of brain serotonin: effect of lithium. *Life Sciences, 9:* 285-290.

Sheard, M.H., Zolovick, A., and Aghajanian, G.K. (1972): Raphe neurons: effect of tricyclic antidepressant drugs. *Brain Research, 43:* 690-694.

Shibuya, T., and Anderson, E.G. (1968): The influence of chronic cord transection on the effects of 5-hydroxtryptophan, l-tryptophan and pargyline on spinal neuronal activity. *Journal of Pharmacology and Experimental Therapeutics, 164:* 185-190.

Shields, P.J., and Eccleston, D. (1972): Effects of electrical stimulation of rat midbrain on 5-hydroxytryptamine synthesis as determined by a sensitive radioisotope method. *Journal of Neurochemistry, 19:* 265-272.

Shopsin, B., Wilk, S., Gershon, S., Davis, K., and Suhl, M. (1973a): Cerebrospinal fluid MHPG. An assessment of norepinephrine metabolism in affective disorders. *Archives of General Psychiatry, 28:* 230-233.

Shopsin, B., Wilk, S., Gershon, S., Roffman, M., and Goldstein, M. (1973b): Collaborative psychopharmacologic studies exploring catecholamine metabolism in psychiatric disorders. In: *Frontiers in Catecholamine Research.* Edited by E. Usdin and S. Snyder. Pergamon Press, New York, Pp. 1173-1179.

Shopsin, B., Gershon, S., Goldstein, M., Friedman, E., and Wilk, S. (1974): Use of synthesis inhibitors in defining a role for biogenic amines during imipramine treatment in depressed patients. *Psychopharmacology Bulletin, 10:* 52.

Shore, P.A., Silver, S.L., and Brodie, B.B. (1955): Interaction of reserpine, serotonin, and lysergic acid diethylamine in brain. *Science, 122:* 284-285.

Shore, P.A., and Brodie, B.B. (1957): LSD-like effects elicited by reserpine in rabbits pretreated with iproniazid. *Proceedings of the Society for Experimental Biology and Medicine (New York), 94:* 433-435.

Sigg, E.B. (1959). Pharmacological studies with Tofranil. *Canadian Psychiatric Association Journal, 4 (suppl):* 75-85.

Sigg, E.B., Soffer, L., and Gyermek, L. (1963): Influence of imipramine and related psychoactive agents on the effect of 5-hydroxytryptamine and catecholamines on the cat nictitating membrane. *Journal of Pharmacology and Experimental Therapeutics, 142:* 13-20.

Sinanan, K., Keatinge, A.M.B., Beckett, P.G.S., and Love, W.C. (1975): Urinary cyclic AMP in "endogenous" and "neurotic" depression. *British Journal of Psychiatry, 126:* 49-55.

Sjoerdsma, A., Engelman, K., Spector, So., and Udenfriend, S. (1965): Inhibition of catecholamine synthesis in man with alpha-methyl-tyrosine, an inhibitor of tyrosine hydroxylase. *Lancet, II:* 1092-1094.

Sjöström, R. (1972): Steady-state levels of probenecid and their relation to acid monoamine metabolites in human cerebrospinal fluid. *Psychopharmacologia, 25:* 96-100.

Sjöström, R., and Roos, B-E. (1972): 5-Hydroxyindoleacetic acid and homovanillic acid in cerebrospinal fluid in manic-depressive psychosis. *European Journal of Clinical Pharmacology, 4:* 170-176.

Sloane, R.B., Hughes, W., and Haust, H.L. (1966): Catecholamine excretion in manic-depressive and schizophrenic psychosis and its relationship in symptomatology, *Canadian Psychiatric Association Journal, 11:* 6-19.

Smith, A.D., and Winkler, H. (1972): Fundamental mechanisms in the release of catecholamines. *Handbook of Experimental Pharmacology, 33:* 538-617.

Snaith, R.P., and McCoubrie, M. (1974): Antihypertensive drugs and depression. *Psychological Medicine, 4:* 393-398.

Snowdon, J., and Braithwaite, R. (1974): Combined antidepressant medication. *British Journal of Psychiatry, 125:* 610-611.

Snyder, S.H. (1972): Catecholamines in the brain as mediators of amphetamine psychosis. *Archives of General Psychiatry, 27:* 169-179.

Sonninen, V., Riekkinen, P., and Rinne, U.K. (1973): Acid monoamine metabolites in cerebrospinal fluid and multiple sclerosis. *Neurology, 23:* 760-763.

Sourkes, T.L., Murphy, G.F., and Chavez, B. (1961): The action of some alpha-methyl and other amino acids on cerebral catecholamines. *Journal of Neurochemistry, 8:* 109-115.

Sourkes, T.L. (1965): The action of α-methyldopa in the brain. *British Medical Bulletin, 21:* 66-69.

Sourkes, T.L. (1973a): Enzymology and sites of action of monoamines in the central nervous system. *Advances in Neurology, 2:* 13-35.

Sourkes, T.L. (1973b): On the origin of homovanillic acid (HVA) in the cerebrospinal fluid. *Journal of Neural Transmission, 34:* 153-157.

Spector, S., Sjoerdsma, A., and Udenfriend, S. (1965): Blockade of endogenous norepinephrine synthesis by alpha-methyl-tyrosine, an inhibitor of tyrosine hydroxylase. *Journal of Pharmacology and Experimental Therapeutics, 147:* 86-95.

Stacey, R.S. (1961): Uptake of 5-hydroxytryptamine by platelets. *British Journal of Pharmacology, 16:* 284-295.

Stein, L., and Ray, O.S. (1960): Acclerated recovery from reserpine depression and monoamine oxidase inhibitors. *Nature, 188:* 1199-1200.

Stein, L. (1962). Effects and interactions of imipramine, chlorpromazine, reserpine and amphetamine on self-stimulation: possible neurophysiological basis of depression. *Recent Advances in Biological Psychiatry, 4:* 288-314.

Stein, L., and Wise, C.D. (1971): Possible etiology of schizophrenia: progressive damage to the noradrenergic reward system by 6-hydroxy-dopamine. *Science, 171:* 1032-1036.

Stitzel, R.E., and Lundborg, P. (1967): Effect of reserpine and monoamine oxidase inhibition on the uptake and subcellular distribution of ³H-noradrenaline. *British Journal of Pharmacology and Chemotherapeutics, 29:* 99-104.

Stjarne, L. (1964): Studies of catecholamine uptake storage and release mechanisms. *Acta Physiologica Scandinavica, 62 (suppl. 228):* 1-97.

Stoof, J.C. (1975): *Neurofarmacologie van m- en p-tyramine.* Proefschrift, Econo-print, Amsterdam.

Strom-Olsen, R., and Weil-Malherbe, H. (1958): Humoral changes in manic-depressive psychosis with particular reference to excretion of catecholamines in urine. *Journal of Mental Science, 104:* 696-704.

Sulser, F., Watts, J., and Brodie, B.B. (1962): On the mechanism of antidepressant action of imipramine-like drugs. *Annals of the New York Academy of Sciences, 96:* 279-288.

Sulser, F., Bickel, M.H., and Brodie, B.B. (1964): The action of desmethylimipramine in counteracting sedation and cholinergic effects of reserpine-like drugs. *Journal of Pharmacology and Experimental Therapeutics, 144:* 321-330.

Sulser, F., Owens, M.L., and Dingell, J.V. (1966): On the mechanism of amphetamine potentiation by desimipramine (DMI). *Life Sciences, 5:* 2005-2010.

Sulser, F., Owens, M.L., Norvich, M.R., and Dingell, J.V. (1968): Relative role of storage and synthesis of brain norepinephrine in psychomotor stimulation evoked by amphetamine or by desimipramine and tetrabenazine. *Psychopharmacologia, 12:* 322-332.

Sulser, F., and Sanders-Bush, E. (1971): Effects of drugs on amines in the CNS. *Annual Review of Pharmacology, 11:* 209-230.

Tagliamonte, A., Taliamonte, P., Perez-Cruet, J., and Gessa, G.L. (1971a): Increase of brain tryptophan caused by drugs which stimulate serotonin synthesis. *Nature/New Biology, 299:* 125-126.

Tagliamonte, A., Tagliamonte, P., Perez-Cruet, J., Stern, S., and Gessa, G.L. (1971b):

Effect of psychotropic drugs on tryptophan concentration in the rat brain. *Journal of Pharmacology and Experimental Therapeutics, 177:* 475-480.

Takahashi, R., Nagao, Y., Tsuchiya, K., Takamizawa, M., and Kobayashi, T. (1968a): Catecholamine metabolism of manic-depressive illness. *Journal of Psychiatric Research, 6:* 185-199.

Takahashi, R., Utena, H., Machiyama, Y., Kurihama, M., Otsuka, T., Nakamura, T., and Konamura, H. (1968b): Tyrosine metabolism in manic-depressive illness. *Life Sciences (part II), 7:* 1219-1231.

Takahashi, S., Kondo, H., Yoshimura, M., and Ochi, Y. (1973): Antidepressant effect of thyrotropin-releasing hormone (TRH) and the plasma thyrotropin levels in depression. *Folia Psychiatric et Neurologica Japonica, 27:* 305-314.

Takahashi, S., Kondo, H., and Kato, N. (1975): Effect of l-5-hydroxytryptophan on brain monoamine metabolism and evaluation of its clinical effects in depressed patients. *Journal of Psychiatric Research, 12:* 177-187.

Tamarkin, N.R., Goodwin, F.K., and Axelrod, J. (1970): Rapid elevation of biogenic amine metabolites in human CSF following probenecid. *Life Sciences, 9:* 1397-1408.

Theobald, W., Büch, O., Kunz, H., Morpurgo, C., Stenger, E.G., and Wilhelmi, G. (1964): Comparative pharmacological studies with Tofranil, Pertofran and Ensidon. *Archives Internationales de Pharmacodynamie et de Thérapie, 148:* 560-596.

Thierry, A-M., Fekete, M., and Glowinski, J. (1968a): Effects of stress on the metabolism of noradrenaline, dopamine and serotonin (5-HT) in the central nervous system of the rat: II. Modifications of serotonin metabolism. *European Journal of Clinical Pharmacology, 4:* 384-389.

Thierry, A-M., Javoy, F., Glowinski, J., and Kety, S.S. (1968b): Effects of stress on the metabolism of norepinephrine, dopamine and serotonin in the central nervous system of the rat: I. Modification of norepinephrine turnover. *Journal of Pharmacology and Experimental Therapeutics, 163:* 163-171.

Thoenen, H., Haefely, W., Gey, K.F., and Hürlimann, A. (1965): Diminished effects of sympathetic nerve stimulation in cats pretreated with disulfiram: liberation of dopamine as sympathetic transmitter. *Life Sciences, 4:* 2033-2038.

Thoenen, H. (1972): Comparison between the effect of neuronal activity and nerve growth factor on the enzymes involved in the synthesis of norepinephrine. *Pharmacology Reviews, 24:* 255-267.

Tissot, R. (1962): Connaissances experimentales sur les monoamines et quelques syndromes psychiatriques. In: *Monoamines et système nerveux central.* Edited by J. de Ajuriaguerra. Georg & Cie, S.A., Geneva. Pp. 169-207.

Todrick, A., and Tait, A.C. (1969): The inhibition of human platelet 5-hydroxytryptamine uptake by tricyclic antidepressive drugs. The relation between structure and potency. *Journal of Pharmacy and Pharmacology, 21:* 751-762.

Toivola, P.T.K., and Gale, C.C. (1972): Stimulation of growth hormone release by microinjection of norepinephrine into hypothalamus of baboons. *Endocrinology, 90:* 895-902.

Trimble, M., Chadwick, D., Reynolds, E.H., and Marsden, C.D. (1975): L-5-hydroxytryptophan and mood. *Lancet, I:* 583.

Tuomisto, J. (1974): A new modification for studying 5-HT antidepressants as uptake inhibitors. *Journal of Pharmacy and Pharmacology, 26:* 92-100.

Turner, W.J., and Merlis, S. (1964): A clinical trial of pargyline and dopa in psychotic subjects. *Diseases of the Nervous System, 25:* 538-541.

Udenfriend, S. (1966): Tyrosine hydroxylase. *Pharmacology reviews, 18:* 43-51.

Ungerstedt, U. (1971a): Stereotaxic mapping of the monoamine pathways in the rat brain. *Acta Physiologica Scandinavica, 367:* 1-48.

Ungerstedt, U. (1971b): Adipsia and aphagia of the 6-hydroxydopamine induced degeneration of the nigro-striatal dopamine system. *Acta Physiologica Scandinavica, 367:* 95-122.

Valzelli, L., Consolo, S. and Morpurgo, C. (1967): Influence of imipramine-like drugs on the metabolism of amphetamine. In: *Antidepressant Drugs*. Edited by S. Garattini and M.N.G. Dukes. Excerpta Medica Foundation, Amsterdam. Pp. 61-69.

Voigtlander, P.F. von, and Moore, K.E. (1970): Behavioural and brain catecholamine depleting actions of V-14, 624, an inhibitor of dopamine-β-hydroxylase. *Proceedings of the Society for Experimental Biology and Medicine, 133:* 817-820.

Wålinder, J., Skott, A., Nagy, A., Carlsson, A., and Roos, B-E. (1975): Potentiation of antidepressant action of clomipramine by tryptophan. *Lancet, I:* 984.

Weir, R.L., Chase, T.N., Ng, L.K.Y., and Kopin, I.J. (1973): 5-hydroxyindoleacetic acid in spinal fluid: relative contribution from brain and spinal cord. *Brain Research, 52:* 409-412.

Weiss, B.L., Kupfer, D.J., Foster, F.G., and Delgado, J. (1974): Psychomotor activity, sleep and biogenic amine metabolites in depression. *Biological Psychiatry, 9:* 45-54.

Weissman, A., Koe, B.K., and Tenen, S.S. (1966): Antiamphetamine effects following inhibition of tyrosine hydroxylase. *Journal of Pharmacology and Experimental Therapeutics, 151:* 339-352.

Werdinius, B. (1967a): Elimination of 3,4-dihydroxyphenylacetic acid from the blood. *Acta Pharmacologica et Toxicologica, 25:* 9-17.

Werdinius, B. (1967b): Effect of probenecid on the levels of monoamine metabolites in the rat brain. *Acta Pharmacologica et Toxicologica, 25:* 18-23.

Westerink, B.H.C., and Korf, J. (1975): Determination of nanogram amounts of homovanillic acid in the central nervous system with a rapid semi-automated fluorometric method. *Biochemical Medicine, 12:* 106-112.

Wharton, R.N., Perel, J.M., Dayton, P.G., and Malitz, S. (1971): A potential clinical use for methylphenidate with tricyclic antidepressants. *American Journal of Psychiatry, 127:* 1619-1625.

Wilk, S., Shopsin, B., Gershon, S., and Suhl, M. (1972): Cerebrospinal fluid levels of MHPG in affective disorders. *Nature, 235:* 440-441.

Wilson, I.C., Prange, A.J., Jr., Lara, P.P., Alltopp, L.B., Stikeleather, R.A , Lipton, M A., and Hill, C. (1973): TRH (lopremone): psychobiological responses of normal women. I. Subjective experiences. *Archives of General Psychiatry, 29:* 15-21.

Winokur, G., Cadoret, R., Dorzab, J., and Baker, M. (1971): Depressive disease. A genetic study. *Archives of General Psychiatry, 24:* 135:144.

Winston, F. (1971): Combined antidepressant therapy. *British Journal of Psychiatry, 118:* 301-304.

Wise, D.C., and Stein, L. (1973): Dopamine-beta-hydroxylase. Deficits in the brains of schizophrenic patients. *Science, 181:* 344-347.

Wong, D.T., Horng, J.S., Bymaster, F.P., Hauser, K.L., and Molloy, B.B. (1974): A selective inhibitor of serotonin uptake: Lilly 110140, 3-(p-trifluoromethylphenoxy)-N-methyl-3-phenylpropylamine. *Life Sciences, 15:* 471-479.

Wooten, G.F., and Cardon, V.P. (1973): Plasma dopamine-β-hydroxylase activity (elevation in man during cold pressor test and exercise). *Archives of Neurology* (Chic.), *28:* 103-106.

Wurtman, R.J., Rose, C.M., Chou, C., and Larin, F.F. (1968): Daily rhythms in the concentrations of various amino acids in human plasma. *New England Journal of Medicine, 279:* 171-175.

Wurtman, R.J., Chou, C., and Rose, C. (1970a): The fate of C^{14}-dihydroxyphenylalanine (C^{14}-dopa) in the whole mouse. *Journal of Pharmacology and Experimental Therapeutics, 174:* 351-356.

Wurtman, R.J., Rose, C.M., Matthyse, S., Stephenson, J., and Baldessarini, R. (1970b): L-dihydroxyphenylalanine: effect on S-adenosylmethionine in brain. *Science, 169:* 395-397.

Wyatt, R.J., Engelman, K., Kupfer, D.J., Fram, D.H., Sjoerdsma, A., and Snyder, F. (1970): Effects of l-tryptophan (a natural sedative) on human sleep. *Lancet, II:* 842-846.

Wyatt, R.J., Portnoy, B., Kupfer, D.J., Snyder, F., and Engelman, K. (1971): Resting plasma catecholamine concentrations in patients with depression and anxiety. *Archives of General Psychiatry, 24:* 65-70.

Yang, H.Y.T., and Neff, N.H. (1973): β-Phenylethylamine: a specific substrate for type B monoamineoxidase of brain. *Journal of Pharmacology and Experimental Therapeutics, 187:* 365-371.

Yang, H.Y.T., and Neff, N.H. (1974): The monoamine oxidases of brain: selective inhibition with drugs and the consequences for the metabolism of the biogenic amines. *Journal of Pharmacology and Experimental Therapeutics, 189:* 733-740.

Youdim, M.B., Collins, G.G., Sandler, M.J., Bevan Jones, A.B., Pare, C.M.B., and Nicholson, W.J. (1972): Human brain: monoamine oxidase, multiple forms and selective inhibitors. *Nature, 236:* 225-228.

Young, S.N., Lal, S., Martin, J.B., Ford, R.M., and Sourkes, T.L. (1973): 5-Hydroxyindoleacetic acid, homovanillic acid and tryptophan levels in CSF above and below a complete block of CSF flow. *Psychiatria, Neurologia, Neurochirurgia* (Amst), *76:* 439-444.

Yuwiler, A., Geller, E., and Eiduson, S. (1959): Studies on 5-hydroxytryptophan decarboxylase. I. In vitro inhibition and substrate interaction. *Archives of Biochemistry, 80:* 162-173.

Zeidenberg, P., Perel. J.M., Kanzler, M., Wharton, R.N., and Malitz, S. (1971): Clinical and metabolic studies with imipramine in man. *American Journal of Psychiatry, 127:* 1321-1326.

Zeller, P., Pletscher, A., Gey, K.F., Gutmann, H., Hegedüs, B., and Strayb, O. (1959): Amino acid and fatty hydrazides: chemistry and action of monoamine oxidase. *Annals of the New York Academy of Sciences, 80:* 555-567.

PART II

Schizophrenic Psychoses

If psychiatry recognizes, at the fundamental level of nosology, that the great problems with which it wrestles, such as schizophrenia or depression, are simply phenomenological clusters with little demonstrated etiological or pathogenetic meaning, it may keep itself from a premature adoption of whatever etiological hypotheses happen to be in vogue, whether they be organic, psychodynamic, or psychosocial. Nosology may then serve best its important purposes of defining and communicating what is known, indicating what is unclear and stimulating the acquisition of evidence in those areas.

S.S. Kety
Classification in Psychiatry and Psychopathology
National Institute of Mental Health, 1968

Introduction

Research strategy in depressions. Tricyclic antidepressants and MAO inhibitors have vigorously stimulated biological depression research in the past 15 years (Van Praag, 1976, Mendels, 1975). Although chemically unrelated, these compounds proved to show two similarities. In psychopathological terms, they have a beneficial effect on (vital) depressions. In biochemical terms, they "mobilize" MA in the brain via different mechanisms. MAO inhibitors inhibit their intraneuronal degradation; tricyclic compounds inhibit the re-uptake of MA into the neuron after transmission of the impulse from one neuron to the other. On these grounds, a correlation is suspected between the therapeutic effect of antidepressants and their MA-potentiating action. This hypothesis received support when it was found that reserpine can provoke typical vital depressions and, on the biochemical level, reduces the amount of MA available in the brain. Reserpine is, so to speak, the counterpart of the antidepressants.

The suspected correlation between the therapeutic and the MA-potentiating action of antidepressants prompted two types of research. To begin with, research focused on the question of whether or not MA are available to a reduced extent in the brain in depressive patients. Secondly, research has been done into the antidepressant potency of substances which more or less selectively potentiate the activity of a particular MA in the brain. The former line of research yielded data suggestive of the presence of a central MA deficiency in certain types of depression. The latter line yielded such substances as chlorampheta-mines, MA precursors and selective "re-uptake inhibitors"—compounds which did not bring about a revolution in the treatment of depressions but did supply useful information on the correlations between MA metabolism, mood and motor activity.

Research strategy in schizophrenia. For many years, up to the end of the sixties, research into the biological substrates of schizophrenia suffered from a lack of more or less well-founded working hypotheses. Without much system, a search was made for abnormal metabolites in a wide variety of body fluids. Strange people, strange substances: a research strategy which was hardly pur-poseful. Recently, it has emerged that the approach which proved to be so productive in depression research is suitable also for psychosis research (Snyder, 1974; Matthyse, 1973; Van Praag, 1975). Like antidepressants, antipsychotic agents such as neuroleptics are a heterogeneous group of compounds in terms of chemical structure with, nevertheless, two levels of similarity. In biochemical terms, they reduce the transmission in DA-ergic and NA-ergic systems in the brain. In psychopathological terms, they exert a beneficial effect on motor unrest and psychotic disturbances in thinking and experiencing. In my view, these facts logically prompt a number of questions which make up an adequate research strategy (Van Praag, 1974, 1975):

(1) Are there indications that human central CA-ergic transmission is like-wise reduced by neuroleptics?

(2) Is there a correlation between reduced CA-ergic transmission and the clinical (side) effects of neuroleptics?

(3) Is the extent to which neuroleptics reduce DA-ergic and NA-ergic trans-mission predictive of their therapeutic action profile?

(4) Does the psychotic patient show signs of increased activity in central CA-ergic systems?

(5) Are compounds which reduce CA-ergic transmission, but do not belong to the group of neuroleptics, therapeutically effective in psychoses?

(6) Do compounds which enhance central CA-ergic transmission have psy-chotogenic properties?

Hypotheses on the pathogenesis of schizophrenia. Research of relevance to the above questions has started only recently. Schizophrenia research lags behind depression research. Yet it has already produced one more or less plausible hypothesis: the so-called "DA hypothesis." This hypothesis postulates that increased activity of central DA-ergic systems, specifically the mesolimbic and mesocortical DA systems, underlies certain features of the schizophrenic syn-drome (Randrup and Munkvad, 1972; Snyder, 1973).

The MA metabolism has been related to schizophrenia in two other ways. One hypothesis postulates hypoactivity of central NA-ergic systems in this syndrome (Stein and Wise, 1971); the other postulates abnormal methylation of 5-HT and DA with synthesis of hallucinogenic metabolites (Osmond and Smythies, 1952; Kety, 1967).

I shall later discuss the DA hypothesis in detail, because it is the most plausible of the three. The other two will be discussed only briefly. Other hypotheses which have come to the fore in the course of the years—for instance, the adrenochrome hypothesis (Hoffer et al., 1954) and the theories on the so-called "pink spot" in urine (Friedhoff and Van Winkle, 1962)—must be dismissed as ungrounded. I shall begin with a chapter on the psychopathology of schizophrenia, outlining the terminological confusion which prevails in this field and briefly discussing my own guidelines in the diagnosis of schizophrenia.

CHAPTER XV
Psychopathological Aspects of Schizophrenic Psychoses

Questions About Schizophrenia

In 1896, in the 5th edition of his *Psychiatrie,* Kraepelin used the designation "dementia praecox" to refer to a group of syndromes which had in common: "a peculiar destruction of internal connections of the personality and a marked damage of emotional and volitional life." This conception encompassed three diagnostic categories of an earlier time: *démence précoce* (Morel, 1860), hebephrenia (Hecker, 1871), and catatonia (Kahlbaum, 1874). Kraepelin differentiated between dementia praecox and a group of syndromes with paranoid symptoms but without severe defects in emotional life and volitional functions, which he called "paraphrenia." In the 6th edition of his textbook (Kraepelin, 1899), he brought dementia praecox and manic-depressive psychosis together under the heading "endogenous psychoses"—a conception which has held its own in the diagnosis of psychoses to this very day.

In his monograph on schizophrenic psychoses, published in 1911, Bleuler rejected the designation dementia praecox because, he maintained, the disease need not inevitably lead to overall deterioration of the personality and does not necessarily start at an early age. He regarded the loss of coherence and interaction between the various psychological functions as the crucial disturbance, and therefore proposed the term "schizophrenia."

The term "schizophrenia" has been generally accepted. Yet the syndrome referred to by this name has remained a persistent source of controversy and uncertainty. The crucial question was and still is: Does schizophrenia exist? In fact, this is a question which has two components. The first is whether schizophrenia is a classical disease entity with a more or less unequivocal "morphology," or, on the other hand, a collective name which covers a whole range of subtypes. This is not really a controversial question, for it is unanimously answered in favor of the latter alternative. The second is whether these subtypes have common characteristics in terms of symptomatology, pathological mechanisms or clinical behavior. If so, they can be brought under a common denominator, and the collective designation of schizophrenia has its justification. If not, the schizophrenia concept has no justification, and there is every reason to view the so-called "subtypes" as independent disease entities, and to study them as such.

There has been no unanimous answer to this problem. Indeed, very few empirical studies have been done on which a conclusion could be based. This has been an enormous handicap in biological schizophrenia research, for we do not know whether the patients in whom this diagnosis has been made are in fact a homogeneous group. The reader should bear this in mind throughout the following chapters. In spite of all this uncertainty, research into biological determinants of schizophrenic behavior has not been unproductive in the past five years. This has been possible because biochemical psychopharmacology has provided such vigorous impulses that the paucity of creative psychopathological research has been compensated.

In the following sections, I shall briefly discuss the symptomatology, causation and course of schizophrenic psychoses—not with the intention of presenting a comprehensive review but primarily in order to demonstrate how unsettled the diagnosis of schizophrenia is, and how sorely this concept needs revision on the basis of empirical psychopathological studies. Finally, I shall describe a simple system of classifying patients with schizophrenic psychoses and selecting, very tentatively, a subgroup which I shall call nuclear schizophrenia. Until more solid diagnostic parameters become available, this type of classification can perhaps be useful in coping with the worst of the prevalent confusion.

Schizophrenic Syndromes

Several different syndromes have been described under the heading "schizophrenia." To begin with, there is *hebephrenia,* an exuberant psychiatric syndrome with massive hallucinations and/or delusions, loss of contact with the environment, and loss of initiative. This syndrome usually starts at an early age, i.e., during puberty or early adolescence. Next comes *paraphrenia,* in which the patient develops systematized but encapsulated paranoid delusions but shows no overall disintegration of the personality (which is why some psychiatrists do not regard this syndrome as schizophrenic). Unlike paraphrenia, *dementia paranoides* is characterized by hardly systematized paranoid delusions and progressive deterioration of the entire personality. The predominant features in *dementia simplex* are loss of initiative and autism, while delusions and hallucinations are not very pronounced. *Catatonia,* finally, is a psychotic syndrome characterized by predominance of motor disorders such as stereotypes, stupor, grimacing, etc.

Not by any stretch of the imagination can these syndromes be brought under the same denominator. They are as dissimilar as circulatory shock and the syndrome of an inflamed gall bladder. But this is not all. Bleuler, for example, has ever expanded the schizophrenia concept. He rejected the independent existence of presenile paranoid states, alcoholic dementia, schizophreniform syndromes associated with brain lesions, infections and intoxications. All these he regarded as latent schizophrenia becoming manifest. Latent schizophrenia, moreover, he believed to be more common than manifest schizophrenia, and with this the schizophrenia concept loses all boundaries. Kraepelin did not show this tendency. On the contrary, he tried to define the schizophrenia concept more and more exactly, though he did not abandon his conception of the many subtypes.

Schizophrenic Symptoms

A second question is whether there exist symptoms characteristic of schizophrenia—symptoms which, regardless of the other components of the syndrome, can be encountered in any type of schizophrenia.

Pathognomonic symptoms. Kraepelin listed the principal symptoms of schizophrenia as poor insight, decline of mental flexibility and performance, blunting of affect, and a loosening of internal unity. As features of more or less secondary diagnostic importance, he listed hallucinations, delusions, depressive mood, motor pathology, volitional disturbances, negativism and "Befehlsautomatie" (automatic response to commands). Bleuler considered loosening of associations and "splitting of the personality" essential for the diagnosis of schizophrenia. Splitting means dissociation: dissociation of emotions from ideas, of expression from emotion, of conduct from intentions, and so on. He regarded these symptoms as direct consequences of a supposed brain disease. He dismissed such symptoms as hallucination, delusion, confusion, twilight states, manic and depressive changes of mood, and catatonic responses as subordinate and not essential to the diagnosis.

Both Kraepelin and Bleuler showed little finesse in making the diagnostic criteria for schizophrenia operational. Moreover, several of the symptoms they deemed characteristic cannot be diagnosed without a fair amount of subjective interpretation, and this impedes their transferability. The most carefully elaborated attempt to describe symptoms pathognomonic for schizophrenia was made by Schneider (1959). He described eleven of them, which he called symptoms of the first rank. The presence of any of these symptoms is sufficient to warrant a diagnosis of schizophrenia, unless gross anatomical abnormalities are present in the brain. The first three are specific types of auditory hallucination:

(1) The patient hears voices that say his thoughts aloud.
(2) The patient hears voices that talk about him.
(3) The patient hears voices that describe his activities.

The fourth symptom is that of delusional percept, that is, a normal perception is followed by an erroneous interpretation with a very special, patient-oriented significance. Symptoms (5) through (11) are expressions of gradual effacement of the boundary between self and environment:

(5) The patient is a passive recipient of exogenous physical influences which he cannot resist; they are associated with physical (hallucinatory) sensations (somatic passivity).
(6) The patient experiences his own thoughts as introduced into him from outside (thought intrusion).
(7) The patient's thoughts are taken away from him by some outside agency (thought removal).
(8) The patient believes that he can transfer his thoughts to others by magic means (thought broadcast).

Symptoms (9) through (11) encompass the patient's belief that his affects, impulses and motor activities are imposed on him by some external agency which also controls him in these respects. These ideas are not accompanied by hallucinatory physical sensations.

Subjective criteria. Schneider's choice of the symptoms of the first order was not based on theoretical considerations, but was entirely pragmatic: they could be most readily diagnosed in a patient. On the other hand, there are psychiatrists whose diagnostic preference is for symptoms which are difficult to objectify, in other words, those which can be established only in communicative contact with the patient. They focus primarily on the patient's ability to enter an affective relationship with another person, and on the extent to which he shows affective "resonance" in such a contact. Disturbances in this ability are regarded as "highly suspect." Conclusions concerning these phenomena are based largely on subjective interpretation, and this is why different evaluators often reach different conclusions. Moreover, the psychiatrist's own ability to establish contact is of course also involved. Last but not least, in actual practice one is often confronted with the questions: Is the patient really suffering from restrictive affect or is he suppressing his feelings? Is this contact really deficient or does he keep at a safe distance due to (neurotic) fear? To put it in diagnostic terms:

Are we dealing with a neurotic reaction with serious repression, or indeed with a schizophrenic syndrome?

Rümke (1967) introduced a purely subjective criterion. He maintained that actually there are no symptoms specific of schizophrenia. The diagnosis is in fact made on the basis of a feeling which the patient evokes in the investigator: the so-called "praecox feeling." He tried to define this as a sense of estrangement: one senses that his own overtures meet an obstacle. The patient accepts nothing of what the investigator offers: contact, warmth, understanding. Recognition of this praecox feeling in himself requires a well-developed empathic ability on the part of the investigator. Since the phenomenon is so difficult to explicitate, it is hardly transferable.

Empirical studies. The Schneider criteria prompt two questions. The first is: Are symptoms of the first rank sufficiently common in schizophrenia to have diagnostic value? The second is: Are they truly pathognomonic—do they differentiate between schizophrenia and other syndromes? Research results in connection with the first question are not unequivocal. Mellor (1970) examined 166 patients diagnosed as schizophrenic for symptoms of the first rank, and found one or several of these symptoms in 119 (72%). In "Schneider-positive" patients, he then studied the frequency distribution of these symptoms. Auditory hallucinations were found in 13% of the patients. Delusional perception and the feeling that emotions and impulses are influenced by others were relatively rare: 6%, 6% and 3% of cases, respectively. The frequency of the other symptoms ranged from 9% to 21%.

Taylor (1972) studied the case histories of 78 patients diagnosed as schizophrenic, and found the above-mentioned symptoms in 22 patients (28%). A study by Carpenter et al. (1973) revealed first-rank symptoms in 51% of 103 schizophrenic patients. Thought removal was the least frequent symptom (15%) and thought broadcast was the most common (33%).

Not only the diagnostic but also the discriminative value of the first-rank symptoms is rather limited. For example, one or several of these symptoms proved to be present also in 23% of patients diagnosed as manic-depressive (Carpenter et al., 1973).

My conclusion is that a wide variety of syndromes have been described under the heading "schizophrenia," and that empirical studies have failed to reveal clearly pathognomonic symptoms. In symptomatological terms, therefore, there are no grounds to bring the various syndromes described as schizophrenic under a common denominator.

Pathogenesis vs. Etiology

Thinking about the origin of schizophrenia has for many years been governed by a controversial question. To formulate it most concisely, this question is whether schizophrenia is a biochemical or a psychosocial disease. Admittedly, the question is by no means specific for schizophrenia. It has been raised in connection with several other psychiatric conditions. Its roots lie in the view that there are two types of disturbed behavior: that in which the brain is disturbed, and that in which it is not. In my view, this dichotomy is unproductive and incorrect (Van Praag and Leijnse, 1963, 1964). It is unproductive because it causes a schism among psychiatrists. On the one side, we find those who expect much, if not everything, from psychosocial intervention. They seek and find the support of more or less radical social reformers. On the other side, we find the protagonists of neurobiology. They have no difficulty at all in feeling at home among their colleagues, the medical specialists.

I have also described the dichotomy as incorrect, for my basic view is that functions of living matter, be they mental or physical, are unthinkable without a material substrate. This implies that behavior (again, regardless of whether it is disturbed or undisturbed) must also have a material (in this case cerebral) substrate. In principle, this substrate can be analyzed. It is the complex of cerebral functional disorders which enables behavior disorders to arise that I describe as the *pathogenesis* of the syndrome. In doing this, I am not advocating an old-fashioned brand of materialism. Numerous factors can contribute to disturbed behavior. Of course, they are not confined to acquired or hereditary imperfections in the cerebral machine, but also include intrapsychic conflicts and tensions in human interrelations. However, these factors, called *etiological* factors, influence behavior not directly, in some sort of vacuum, but via functional changes in the brain—changes which jointly make up the cerebral substrate underlying the behavior disorders. In the case of schizophrenia, biochemical changes in the brain enable this complex of behavior disorders to occur. In this sense, schizophrenia is a biochemical disease. Another question is how schizophrenic behavior disorders, including the cerebral substrate which generates them, are caused: chiefly by psychosocial factors, by somatic factors (acquired or hereditary), or by a combination of these. This is a question of a different order and open to a difference of opinion.

The question whether schizophrenia is a biochemical or rather a psychosocial disease is of the same order as the question whether pneumonia is a disease caused by infiltrates in the lung or by pneumococci. I mean simply that this question poses a spurious problem, for it presents two possible causative factors with the invitation to choose one. The two factors presented, however, are not alternatives but related links in a chain of causes and effects.

My remarks in the three following sections pertain exclusively to the etiology of schizophrenic psychoses. Possible pathogenetic factors are not discussed. They are the quintessence of the subsequent chapters.

Etiological Views: Hereditary Factors

Kraepelin included dementia praecox among the chiefly hereditary psychoses. Genetic research in subsequent years has demonstrated that hereditary factors are indeed involved (reviews by Rosenthal, 1970; Strömgren, 1975; Kety, 1975). The arguments derive from three types of research. The first is family studies. Schizophrenia is much more frequent among close relatives of schizophrenic patients than in the total population, and the frequency is as much higher as the blood relationship is closer, i.e., higher in brothers, sisters, children and parents of schizophrenic patients than in uncles, aunts, nephews and nieces. In second-degree relatives, too, the incidence is higher than that in the total population (if not much higher). The marked accumulation of schizophrenic pathology in certain families has also been used as an argument by protagonists of family theories on schizophrenia. They propose that, rather than parent-child transmission of genes, family pathology is responsible for the disease, that is, the chaotic, confusing climate in which the child is brought up and in which it learns irrational ways of reacting.

The second type of research involves twin studies. The concordance rate of schizophrenic psychoses is substantially higher in monozygotic than in dizygotic twins. This phenomenon, too, need not necessarily be interpreted genetically. The ties between identical twins are often very close, and the tendency to identify is very strong. If one of the twins becomes psychotic, then the other is very likely to develop a similar pattern of behavior. This is a psychological interpretation which is not entirely plausible. In identical twins brought up separately, the concordance rate of schizophrenia was likewise high, and comparable with that in twins brought up together.

The third type of research involves studies of children born from one or two schizophrenic parents but brought up, not by the biological parents, but in foster families. So far, all these studies have led to the same conclusion. The incidence of schizophrenic psychoses was much higher in these children than in the children of their foster parents, with whom they were jointly brought up.

Transmission is probably not by a single gene but by a number of genes, each of which adds a given element, or dimension, to the schizophrenic behavior pattern. The variability of schizophrenic symptomatology; the often irrefutable significance of environmental influences; the frequent impossibility of establishing or excluding the diagnosis with certainty; the gradual transition between what is called normal if somewhat eccentric behavior and schizophrenic behavior—all these factors are more readily explained by a polygenic theory than by a theory which postulates a single active gene. In genetic terms, a variable number of pathological genes carried could readily explain the variability of symptoms and clinical courses within the group of schizophrenic psychoses.

Etiological Views: Exogenous Factors

Schizophrenia, however, has proved not to be linked exclusively to endogenous factors. The syndrome was observed also in patients with demonstrable anatomical brain damage (review by Reid, 1973). The widely discussed question has been whether these cases involved "genuine" schizophrenia, which had been latent prior to the brain injury and was made manifest by it, or really organic schizophrenia. Since the concept "latent schizophrenia" has hardly been operationalized (although it is certainly multi-interpretable), and empirical studies were consequently omitted, the question was never answered and continued to serve as an appreciated subject of learned discussions.

On the basis of my own clinical, but not systematized experience, I would say that the label "provoked latent schizophrenia" is unsuitable in quite a few cases of schizophrenia associated with organic cerebral lesions. I reach this conclusion in view of an untainted family history, a "clean" psychiatric history and a more or less normal premorbid personality structure in these cases. On this basis, I suspect that organic schizophrenia does exist. An argument against my suspicion is that the anatomical lesion in organic schizophrenia has no specific characteristics as to type or localization. It can be of an inflammatory, degenerative or neoplastic type, and, so far as we know, schizophrenic syndromes can be observed in association with virtually any localization. Given a direct causal relationship, one would expect schizophrenic behavior disorders to occur only in association with particular lesions of a particular localization. What can be concluded is that the problem of organic schizophrenia is yet to be solved.

The term "organic" invites a few additional remarks in passing. In psychiatry, the term is used with reference to a group of psychoses which occur in the presence of a demonstrable anatomical brain lesion. The psychoses in which this is not the case are sometimes referred to as "functional." This has gradually led to the misconception that only organic psychoses have a cerebral substrate. I have pointed out the untenability of this view above. If this terminology is to be maintained, then the two types of psychosis should be defined as follows. Organic psychoses are psychoses in whose etiology anatomical brain damage plays an important role. Their pathogenesis, i.e., the complex of cerebral functional disorders which enables the psychotic behavior disorders to occur, is unknown. In functional psychoses there are no demonstrable anatomical lesions in the brain; of their pathogenesis we possibly know a few features (to be discussed in subsequent chapters), and in their etiology the emphasis seems to be on hereditary and/or psychosocial factors.

Etiological Views: Psychological and Social Factors

Bleuler's "reactive schizophrenia" and what this conception has brought about. More frequently than anatomical lesions of the brain, disorders of the (premorbid) personality structure were observed in patients with schizophrenic symptoms, and the period preceding the manifest psychosis was found to have been colored by intrapsychic and/or more relationally determined tensions. In the 4th edition of his textbook (1923), Bleuler differentiated between "process schizophrenia" and "reactive schizophrenia"—forms which overlap in symptomatological terms but differ in that the former should have a chiefly organic etiology, whereas the latter should be largely psychogenic. This distinction has given rise to numerous obscurities in classification, symptomatological hair-splitting, and etiological misunderstandings which have continued to burden psychiatric diagnosis to this very day. I will discuss only a few, since any attempt at comprehensiveness would produce a whole book on the evolution of diagnostic thinking in psychiatry.

(1) Many followed Bleuler's example and divided the schizophrenias into two groups. As to nomenclature, everyone followed his own tast. This led to such categories as true schizophrenia vs. schizophreniform psychoses (Langfeldt, 1939), true schizophrenia vs. pseudo-schizophrenia (Rümke, 1967), and many others. As the number of dichotomies increased, the criteria were watered down. The dichotomy was made on the basis of etiology (psychogenic vs. nonpsychogenic), course (poor vs. good prognosis), and symptomatology (presence or absence of so-called "essential symptoms"), as such or in various combinations.

(2) New categories of psychoses were distinguished and differentiated from schizophrenia. By way of example, I mention psychogenic and degenerative psychoses (Faergeman, 1963; Van Dijk, 1963). Psychogenic psychosis was differentiated on the basis of etiological considerations. This category encompassed psychoses interpreted as the catastrophic nadir of a neurotic development—as the extreme consequence of unsolved psychological problems. The differentiation of a category of degenerative psychoses was justified with reference to symptomatology, etiology and course. This type of psychosis was described as one with frequent relapses, without a distinct psychogenic cause, and complete remission to the premorbid level after each phase. Moreover, the syndrome varied markedly from phase to phase. In recurrent forms of schizophrenia, the syndrome was believed to be much less variable.

How valid are these differentiations? For example, is the course of a psychogenic psychosis essentially changed by psychotherapy, or does the syndrome tend to relapse nevertheless, the psychogenic provocation becoming less and less evident? Such a course of events is repeatedly seen in "true" schizophrenia. Is it really true that the personality structure remains undamaged in degenerative psychosis, or do these patients come to resemble chronic schizophrenics after all? In the absence of empirical studies, these questions must remain moot.

In any case, genetic findings do not justify these differentiations. Conditions known as schizophreniform psychosis, symptomatic schizophrenia, psychogenic psychosis, degenerative psychosis, atypical or reactive schizophrenia, schizoaffective psychosis, or borderline schizophrenia are all genetically linked to "true" schizophrenia. They seem to be members of the same genetic family (Rosenthal, 1970).

(3) The introduction of the psychogenic schizophrenia concept also caused partial effacement of the boundaries between schizophrenia and nonpsychotic

behavior disorders. A wide variety of strange, inexplicable character traits; a tendency toward reticence; relational deficiencies (to mention only a few phenomena): these were all labeled "schizoid" or, worse, were diagnosed as pseudoneurotic schizophrenia and given a room in the vast mansion called schizophrenia (Reich, 1975). This is one of the reasons for the fact that statistics on the incidence of schizophrenia in different countries and institutes show differences which ridicule the concept of scientific classification (Kendell, 1975; Falek and Moser, 1975).

So much for the diagnostic confusion caused by the term reactive (Psychogenic) schizophrenia. Let us now consider the concept itself.

The etiological weight of psychosocial factors. Few psychiatrists will maintain that the psychopathology of the schizophrenic patient as a rule begins with the manifestation of the psychosis. In many cases, the life history reveals such factors as an unhappy parental marital life; a youth during which the patient was exposed to excessive expressions of aggressivity and hostility among the parents; abnormalities in the personality of one or both parents; and so on (review by Arieti, 1974). Family theorists such as Lidz, Bateson, Laing and many others (review by Sedgwick, 1975) have rightly emphasized that mutual relations in the home of a schizophrenic patient can be very disturbed, "schizophrenic." There is no question about this. Questions arise when we consider the etiological weight of these factors. Can they constitute a principal cause, or is their significance limited to provocation of latent schizophrenia? As with organic schizophrenia, an unequivocal answer to this question cannot be given. On the one hand, every clinician regularly sees cases in which psychogenic problems are inextricably entangled in the manifestation and in the content of the psychosis. This is an argument in favor of the psychogenic schizophrenia concept. An argument against this concept is the absence of any specificity. Psychodynamic mechanisms described in schizophrenia are also encountered in a variety of nonschizophrenic, nonpsychotic psychiatric disorders.

The same applies to family tragedies. There is not a trace of evidence that the communication pathology described is specific to the family life of schizophrenic patients, or that this pathology accumulates only within this category of patients (editorial, *Lancet,* 1975). Authors in this corner of the field, moreover, tend to think somewhat naively in analogs. They discern a direct causal relation between irrationalities which they discover in the patient's family relations, and irrationalities in the patient himself. They conclude that the disturbed behavior, e.g., the patient's conviction that his thoughts are literally removed from him, is directly provoked by the domineering mother; his incoherent line of thought is derived, or rather learned, from the sado-masochistic associations between his parents. For the sake of convenience, it is overlooked that the mother is not suffering from delusions and that the parents are capable of coherent communication (unless they suffer from schizophrenic psychoses). This is reminiscent of an attempt to explain the excretory functions of the body directly from the ingestion of food, ignoring such factors as digestion and metabolism. Moreover, family theorists have tended to disregard the notion that schizophrenic family tragedies could develop secondarily, i.e., in response to the patient's disturbed (and disturbing) behavior.

Be this as it may, the aspecificity of the psychodynamic and psychosocial mechanisms reduces their etiological weight and necessitates introduction of a factor x, which renders the psychosis manifest. This is why of the two concepts—psychogenically provoked schizophrenia and psychogenic schizophrenia—the former seems to be the most plausible.

To avoid misunderstandings, it should be pointed out that my use of the word "aspecificity" with regard to the psychodynamic mechanisms described, refers to the *form* of the disturbed behavior. When the same mechanism can lead to, say, a delusional syndrome, a hallucinatory syndrome or a compulsive syndrome, then it is inadequate as a mode of explanation. A different matter is that this mechanism can indeed be specific with regard to the *content* of the disturbed behavior. A delusion of sin, the hearing of threatening voices, and a washing ritual can each as such give expression to unassimilated guilt feelings rooted in the individual course of development. Very often, psychodynamic and psychosocial mechanisms do indeed have a high degree of specificity with regard to the substance of disturbed thinking and experiencing, and they can give a modicum of insight into the sense of the psychotic "nonsense."

Etiological Views: Conclusion

The etiology of schizophrenic psychoses involves, not a single factor, but a pattern of factors: so much seems certain. However, we do not know what this pattern looks like, nor whether one pattern underlies all types of schizophrenia or whether a given pattern of factors is required for a given syndrome to become manifest. For the time being, therefore, there is no reason to bring schizophrenic syndromes under a common denominator on the basis of etiological criteria.

The Course of Schizophrenic Psychoses

In Kraepelin's definition of dementia praecox, the poor prognosis was a central feature. The patient would either show gradual deterioration without evident remissions, or an intermittent course with incomplete remission after each phase, finally leading to a chronic defect state. Let me mention in passing that the diagnosis "chronic schizophrenia" is very ill-defined. In actual practice, the designation serves as a receptacle for numerous serious but possibly diverse personality defects which markedly impede the patient's social adjustment and in which there is no demonstrable interference with the anatomical integrity of the brain.

In his monograph, Bleuler made a rather half-hearted effort to attack the prognostic criterion. Somewhere in the text, he stated that schizophrenia is incurable, or at least that *restitutio ad integrum* does not occur, but elsewhere he maintained that the disease can recede or be arrested in any stage.

In subsequent years, the prognostic criterion has been used entirely at will. Some authors have remained faithful to Kraepelin, using the term "schizophrenia" only if the prognosis is obviously unfavorable and considering complete recovery to be inconsistent per definition with a diagnosis of schizophrenia. In the case of complete recovery, they speak of schizophreniform psychosis or pseudo-schizophrenia or classify the syndrome in a separate diagnostic category, e.g., that of psychogenic or degenerative psychoses. Others regard the prognostic criterion as inconclusive and diagnose schizophrenia mainly on symptomatological grounds. They belong to the group of psychiatrists who have claimed therapeutic, especially psychotherapeutic success, and who vehemently refute the incurability of schizophrenia. This may seem to be a battle of principles; in reality, it is a semantic conflict. If we define the term "carcinoma" as a collective name for all (benign and malignant) tumors, we can claim a high cure rate; if we confine the term "carcinoma" to malignant tumors, the cure rate is low. Let me add immediately, however, that the emphasis on the curability of schizophrenia has been beneficial in a practical sense: types of schizophrenia with, in principle, a favorable prognosis received adequate (psycho)therapy, and schizophrenic patients bound to show an unfavorable prognosis were treated more humanely than in the past.

The controversy intensified with the introduction of the neuroleptics. Many patients regarded as schizophrenic made a rapid recovery in response to these agents. Is this a reason to revise the diagnosis, or are there two types of schizophrenia—one that is curable and the other that is not (yet) curable? The answer to this question is of necessity an arbitrary one. Our present knowledge warrants no well-argued decision. We are waiting for a decision *ex cathedra* by a psychiatric council, or for empirical studies. I definitely prefer the latter.

In all, we find that the syndromes called "schizophrenic" can apparently take any conceivable course: rapid complete recovery; recovery with residual personality damage; a chronic recurrent course with or without personality damage after each active phase; and a chronic course. The prognostic criterion, therefore, is unsuitable to unite all the syndromes characterized as schizophrenic.

Three-Dimensional Classification of Schizophrenic Psychoses

Etiological aspecificity of psychiatric syndromes. In psychopathology, there is no predictable relation between etiology and syndrome. For example, psychogenic factors can be responsible for neurotic symptoms involved in the development of psychoses, and causative of dissocial behavior. Inversely, it is impossible to infer the etiology from the nature and course of a given syndrome. For example, a vital depression can be of chiefly hereditary determination; it can occur in response to severe psychosocial stress; or it can be provoked by somatic diseases such as a viral infection like infectious hepatitis.

The fact that psychiatric syndromes are etiologically aspecific has an important implication. The implication is that these behavior disorders must always be classified on the basis of three criteria: *symptomatology, etiology* and *course.* This means that one-word diagnoses such as neurosis, depression and schizophrenia are futile. There are two possibilities: either they suggest something about symptomatology, etiology and course simultaneously, but not explicitly, giving these concepts the diagnostic depth of focus of a poor photograph (e.g., the schizophrenia concept), or they are used sometimes in the symptomatological and at other times in the etiological sense, thus creating chaos rather than diagnostic order (e.g., the neurosis concept).

In the following paragraphs, I shall apply the principle of three-dimensional classification to the group of psychoses which generally involve no clouding of consciousness, in which there is usually no severe anatomical brain damage, and which are commonly referred to as "schizophrenic." This procedure makes it possible to form groups of patients which tend to be homogeneous and are recognizable to other investigators. These are prerequisites for reduplication studies. Moreover, this is the only way to distinguish, within the group of schizophrenic psychoses, subgroups which differ from other categories in, say, response to therapy, pathogenesis or prognosis. This method of classification is primarily intended for research purposes, but I believe that it is useful also in daily practice and enhances the precision of diagnostic evaluation.

Symptomatological typology of schizophrenic psychoses. A system of diagnostic classification is not a catalog. It encompasses not all conceivable symptoms but only those which are often prominent. In the classification of each individual patient, I propose to mention which of the following symptoms are present.

(1) Lowered level of consciousness. The old rule that consciousness is unclouded in schizophrenia seems to be generally correct, although in acute psychoses which occur in response to marked psychological stress, the level of consciousness can be lowered during the first few days or weeks. In the more chronic schizophrenic psychoses, consciousness is always unclouded.

(2) Delusions, defined as ideas whose content is probably incompatible with reality but which nevertheless cannot be corrected. The content of these ideas is indicated whenever paranoid delusions or delusions of influence are involved. The former category comprises ideas of being in some way wronged by external

agencies; the latter, ideas that external agencies influence one's thinking, feeling or physical functioning, or that one has a similar power to influence others. The suggestion to note specifically whether a delusion is paranoid arises from the uncertainty as whether paranoid states do or do not come under the heading "schizophrenia." This uncertainty is upheld by two data: the fact that the personality structure outside the paranoid foci is often quite intact, and the fact that the course is often not progressive. The suggestion to note specifically whether a delusion is one of being influenced is based simply on the fact that this phenomenon has of old been regarded as "typically schizophrenic."

(3) Hallucination and delusional percept. A true hallucination involves sensory perception without sensory stimulation. If there is sensory stimulation but evidently faulty interpretation (e.g., the patient regards a glass bowl in his room as a functioning television screen), then the term "delusional percept" applies. In schizophrenic psychoses, auditory and tactile hallucinations (illusions) are believed to be predominant; the latter are often associated with delusions of being influenced. In organic psychoses, visual hallucinations are believed to be predominant.

(4) Emotional flattening. A certain chilliness characterizes contacts with the patient. There is little evidence of (adequate) emotions at appropriate moments. The investigator consequently has a sense of difficulty in establishing a relationship with the patient. There is no affective rapport. Attempts at empathy are unsuccessful. This emotional flattening or chilliness is most apparent in schizophrenic psychoses with a chronic course; it is by no means always pronounced in acute forms.

(5) Motor activity. Schizophrenic psychoses can be accompanied by motor retardation, to the point of stupor, but also with hyperkinesia: grimacing, stereotyped movements, strange movements, etc. In the so-called catatonic stupor, there is often more active resistance than passive endurance; this is evident from the high muscle tonus and the resistance encountered when an attempt is made to move, say, the limbs. The abnormal movements often have a symbolic significance for the patient.

(6) Inertia. Loss of initiative is common in schizophrenic psychoses. The patient can hardly bring himself to meaningful activities. He stops working, stays at home, spends much time in bed. Contacts with the environment become scanty. Sometimes the inertia disappears as the psychosis disappears; in other cases, it persists, or even increases, after the disappearance of the other symptoms. In this case, we have a cure with a residual defect. Inertia is an ominous symptom because it is so difficult to treat, and it can render the patient quite unsuitable for an existence in normal society.

(7) Incoherent train of thought. By no means in all types of schizophrenic psychosis is this symptom noticed. In association with unclouded consciousness, it signifies severe, diffuse disintegration.

Etiological typology of schizophrenic psychoses. A carefully taken family history gives an impression of the weight of the endogenous (hereditary) factor. This factor is considered to be positive if psychoses are or have been present in first-degree or second-degree relatives. I propose that it be regarded as not positive if the taint exclusively involves manic-depressive manifestations, for there are arguments which indicate that schizophrenic psychoses and the group of manic-depressive syndromes are genetically different. Manic-depressive manifestations are rare in families with schizophrenic psychoses, and vice versa. Moreover, there has been no report on monozygotic twins of whom one was regarded as schizophrenic while the other suffered from a manic-depressive syndrome (Rosenthal, 1970).

Involvement or noninvolvement of pathogenic factors of a physical nature can be established by internal examination and neurological examination. The diagnosis should specify whether a cerebral lesion has been demonstrated, or a systemic disease which can be assumed possibly to interfere with cerebral function.

Finally, the contribution of psychogenic and social factors is to be evaluated. Here lurk dangers of deflation as well as of inflation. It is as confusing to inflate events of little significance to severe psychotraumata as it is to ignore serious frustrations or to wave them away as insignificant. Psychological stress prior to manifestation of the psychosis does not as such warrant the use of the term "psychogenic" (unless the stress has been excessive, and hardly bearable for a majority of people, e.g., the violence of war, natural disasters, etc.). It should be made plausible why this stress (which is usually assimilated independently) has exceeded the coping capacity of this particular individual. This is the case if:

(1) A chronic intrapsychic conflict situation can be traced, which the patient cannot solve unaided and on which the stressful life situation was directly imposed.

(2) The patient's (premorbid) personality structure shows weak spots which explain why he was unable independently to solve the conflict.

(3) The weak spots in the personality structure can be related to disturbed psychological development.

The use in actual practice of the terms "psychogenic" and "idiopathic" as synonyms is a serious devaluation of the former concept.

Etiologically, therefore, a given psychosis is classified as being of idiopathic, hereditary (endogenous), exogenous or psycho(socio)genic origin, or determined by a combination of the last three factors.

Typology of schizophrenic psychoses according to course. On the basis of this criterion, the following categories can be distinguished.

(1) First psychotic phase with a duration of less than one year.

(2) A history of several psychotic phases, each followed by restoration to the premorbid level. The number of phases to be stated in the classification.

(3) A history of several psychotic phases, with increasing damage to the personality structure after each phase.

(4) Chronic schizophrenic psychoses: continuously existing for a year or longer despite more or less adequate treatment (medication, psycho(socio) therapy).

(5) Like (4), but with no or with evidently inadquate treatment. The duration of the psychosis to be stated for categories (4) and (5).

Proposed nomenclature. I regard psychoses which take their course without clouding of consciousness as schizophrenic psychoses. The diagnosis is made exclusively on symptomatological grounds: presence of psychotic symptoms, of whatever type, while consciousness remains unclouded. The diagnosis is qualified as follows: schizophrenic psychosis characterized by . . . (follows an outline of the principal symptoms), provoked by . . . (follows an outline of the principal etiological factors) and with a . . . (follows specification) course.

I deliberately use the term "schizophrenic psychosis" rather than "schizophreniform (schizophrenia-like) psychosis," because there is no agreement on what constitutes "true" schizophrenia. I do want to maintain the concept behind the designation "true schizophrenia," but on the basis of the considerations already presented I prefer the designation "nuclear schizophrenia." I use the course as sole criterion for this diagnosis, describing a schizophrenic psychosis as nuclear if its course is unfavorable: chronic or chronic recurrent with personality deterioration.

The deterioration should occur in spite of therapy which can be considered adequate, with both pharmacological and nonpharmacological means. Nuclear schizophrenia is symptomatologically a schizophrenic psychosis because consciousness remains unclouded during its course. Otherwise, this concept has been entirely detached from symptomatological considerations and from the factor etiology. I propose this unlinking because neither clinical features nor premorbid personality structure or presence or absence of precipitating factors warrants a reliable prediction about the prognosis of a schizophrenic psychosis (Serban and Gidynski, 1975).

Tentatively, I should like to circumscribe one group within the schizophrenic psychoses: that of the psychogenic psychoses. Apart from the fact that a disturbance of consciousness *may* occur during the first days or weeks, this category does not fundamentally differ symptomatologically from that of the schizophrenic psychoses. The diagnosis is based mainly on etiological grounds: (a) evident involvement of psychosocial factors; (b) fading of the syndrome soon after removal of the patient from the situation which has become unbearable; (c) emergence of a neurotic personality structure after recession of the psychosis; (d) the investigator has reason to expect that psychotherapy will lead to stabilization of the personality and so to reduction of the risk of a relapse.

The psychogenic psychosis concept does not coincide with the category described by Robins and Guze (1970) as schizophrenia with a favorable prognosis. Criteria for the latter category include: acute onset, relatively undisturbed premorbid personality structure, and a family history which is untainted or tainted only by manic-depressive manifestations. In the group of psychogenic psychoses, however, neurotic problems of adjustment in the premorbid life situation are the rule, and the family history does include psychoses. Cases in which disintegration is gradual, over several weeks, are not uncommon.

Psychogenic psychosis is a provisional diagnosis because, not infrequently, the psychosis shows a relapse tendency despite psychotherapy, and in subsequent psychotic phases the psychogenic etiology is less and less apparent, leading to the conclusion that we are after all dealing with a schizophrenic psychosis.

I do not by any means propose this classification system as definitive. It is merely an aid, to be used in an effort finally to find a way out of the labyrinth we call schizophrenia. And on the way it can be adjusted, if necessary.

Conclusions

The schizophrenia controversy boils down to two questions. The first is: Does schizophrenia exist as a recognizable disease entity? The second is: Is schizophrenia a biochemically, psychologically or socially determined disease? In other words: Is its main cause to be found in intracerebral metabolic disorders, in intrapsychic conflict situations, or in chronically disturbed family relations?

Kraepelin presented a fairly rigid definition of schizophrenia (although he called it dementia praecox). It was a syndrome with a given symptomatology, of chiefly hereditary determination, and with a poor prognosis. The schizophrenia concept has been more and more loosely used since Kraepelin. Schizophrenia, it was believed, could be caused also by psychogenic factors and by anatomical brain lesions. The prognosis need not be unfavorable; recovery was believed possible. The specificity of the symptoms became a perpetual source of dispute. Which symptoms are characteristic, which are secondary? Empirical studies were not started until half a century after Kraepelin's time, and have so far failed to reveal any pathognomonic symptoms, i.e., symptoms which are characteristic of schizophrenia and only sporadically found in other syndromes. For the time being, neither clinical features nor etiology nor course justifies attempts to bring all the syndromes described as schizophrenic under a common denominator. This means that "schizophrenia" is a completely inadequate diagnosis. One should diagnose a schizophrenic psychosis with a given symptomatology, a given etiology and a given course. The term "schizophrenic psychosis" is simply a collective name for psychoses which take their course without clouding of consciousness. The available empirical data do not permit a more exact definition. This is why a three-dimensional diagnosis of schizophrenic psychoses—their consistent classification according to three criteria: symptomatology, etiology and course—is an indispensable requirement if ever we are to find our way out of the labyrinth of the schizophrenic psychoses. In any case, this three-dimensional approach would benefit psychiatric diagnosis in general. In the following chapters, the term "schizophrenia" will nevertheless be encountered, simply because it is being used in the literature I shall be discussing.

Within the group of the schizophrenic psychoses, I should like to reserve the designation "nuclear schizophrenia" for syndromes which take an unfavorable course and are associated with gradually progressive deterioration of the personality. I propose to accept the course as sole cirterion and to detach the designation from symptomatological and etiological considerations.

The question whether "schizophrenia" is a biochemical or rather a psychosocial disease poses a spurious problem. Behavior is unthinkable without corresponding cerebral substrate. Disturbed behavior therefore presupposes a disturbed cerebral substrate. At least in principle, this disturbed cerebral substrate can be described in biochemical terms. In this sense, the schizophrenic psychoses, like any other behavior disorders, are biochemical diseases. A different question is whether the behavior disorders, including their cerebral substrate, are determined chiefly by psychosocial or mostly by (acquired or hereditary) somatic factors. It seems likely that both groups of factors play a role, but in varying relations of importance.

As long as pathogenesis and etiology are not consistently differentiated in psychiatry, biologically and psychosocially orientated, psychiatrists will face each other across a chasm, and consequently theorization about and research into the causes of disturbed behavior will stagnate.

CHAPTER XVI

Neuroleptics and CA-ergic Transmission in the Brain
Animal Experiments

Neuroleptics and CA Turnover

The CA turnover is increased by neuroleptics. The CA have had a central place in studies of the mode of action of neuroleptics. The reason is that all neuroleptics have the ability to provoke hypokinetic-rigid symptoms in human individuals, and in the course of the Sixties it was established that this syndrome, so far as it develops in Parkinson's disease, is related to a DA deficiency in the basal ganglia. This raised the suspicion that at least the DA metabolism is influenced by neuroleptics.

When an animal is given a neuroleptic, the central CA concentration remains unchanged. Yet the metabolism of the CA changes, as demonstrated by the fact that the concentration of CA metabolites increases (Carlsson and Lindqvist, 1963; Andén et al., 1964). This applies to 3-methoxytyramine as well as to normetanephrine, methylated degradation products of DA and NA, respectively, and also to the acid DA metabolite homovanillic acid (HVA). The concentration of 5-HIAA, the principal degradation product of 5-HT, remains unchanged (Andén et al., 1964). The neuroleptics are a chemically heterogeneous group, which includes such diverse compounds as phenothiazines (e.g., chlorpromazine: Thorazine), butyrophenones (e.g., haloperidol: Haldol), and diphenylbutyl-piperidines (e.g., pimozide: Orap). The above described effect, however, is peculiar to all these neuroleptics, regardless of their chemical structure.

The combination of an increased metabolite concentration with an unchanged concentration of the mother amines suggests an increased degradation of CA the loss being compensated for by an increased synthesis. In other words, the combination is indicative of an increased CA turnover. This suspicion is substantiated in two ways. To begin with, in isotope studies, when tyrosine, the mother substance of the CA, is radioactively labeled and then injected into test animals, the CA in the brain are labeled. The rate at which the label is incorporated into and disappears from the CA, is increased by neuroleptics. Incorporation and disappearance of the label can be read as: rate of synthesis and degradation of CA (Nybäck and Sedvall, 1970b).

Another method is that which utilizes enzyme inhibitors. Inhibition of tyrosine hydroxylase, the enzyme which catalyzes the first step in CA synthesis, causes loss of DA and NA from the brain. Inhibition of DA-β-hydroxylase, the enzyme which changes DA to NA, causes loss of NA only. In both cases, the rate of CA disappearance is increased by neuroleptics. This suggests that they stimulate CA degradation (Andén et al., 1972).

Possible causes of the increased CA turnover. Increased CA turnover with unchanged CA concentration suggests an increased CA synthesis, which in turn indicates activation of the enzyme tyrosine hydroxylase, for this is the factor which limits the speed of this process. *In vitro,* neuroleptics exert no influence on tyrosine hydroxylase (Pletscher et al., 1967). Activation of this enzyme *in vivo* therefore probably takes an indirect course. There are two possible hypotheses on the underlying mechanism (Bartholini and Pletscher, 1972).

The first hypothesis is based on the observation that neuroleptics, at least those of the phenothiazine type, can reduce the uptake of CA from the synaptic cleft into the neuron (Thoenen et al., 1965; Pollard et al., 1975). This can be assumed to lead to (a) reduced concentration of cytoplasmic CA; (b) increased concentration of CA at the postsynaptic receptors, and therefore activation of these receptors. It is possible that activation of tyrosine hydroxylase is based on one of these two processes, but it is not likely.

The second hypothesis assumes that neuroleptics block CA receptors and therefore reduce transmission in CA-ergic neurons (Van Rossum, 1967). Via feedback mechanisms, this is believed to lead to an increased flow of impulses in the presynaptic element, and this in turn to activation of tyrosine hydroxylase and increased CA production. The increased CA production is thus viewed as some sort of compensatory mechanism to overcome the receptor block. With the aid of the technique of single-unit recording, it has been demonstrated that neuroleptics do increase the activity in DA-ergic cells (Bunney et al., 1973). Nature and localization of this supposed feedback mechanism, however, are not precisely known. In principle, it could be localized at a presynaptic or postsynaptic site. In the former case, one might think of DA-sensitive receptors in the presynaptic membrane, which regulate the rate of synthesis on the basis of the amount of DA in the synaptic cleft. In that case, we must assume that these receptors, like the postsynaptic receptors, are blocked by neuroleptics; their decreasing state of excitation would then be a signal for an increase in DA synthesis. With regard to a postsynaptic feedback mechanism, one could think of a chemical compound which diffuses through the synaptic cleft or of a nerve pathway. No such nerve pathway has been demonstrated. Andén and Bédard (1971) and Bartholini and Pletscher (1971) have suggested that this could involve cholinergic fibers, for they found that anticholinergics such as atropine antagonize the HVA accumulation in the corpus striatum which is provoked by neuroleptics. However, this phenomenon was not confirmed by Westerink and Korf (1975a).

Notes on the Theory that Neuroleptics Block CA Receptors

Of the two above-mentioned theories, the latter (that of receptor block) is more plausible than the former. To begin with, peripheral DA receptors have been demonstrated to be blocked by neuroleptics (Woodruff, 1971). There are indications, moreover, that stimulation of postsynaptic CA receptors (e.g., as a result of uptake inhibition) reduces rather than increases the activity of tyrosine hydroxylase (Costa and Neff, 1966; Alousi and Weiner, 1966). Finally, one would expect on the basis of the former hypothesis that neuroleptics increase CA-ergic activity in the brain, and there are no indications that they do.

The latter hypothesis implies that neuroleptics reduce CA-ergic activity. This is plausible for two reasons: neuroleptics arrest the effects of CA-ergic hyperactivity, and they induce symptoms of CA-ergic hypoactivity. Let me present a few elucidating examples.

Neuroleptics as CA antagonists. There are several methods of increasing CA-ergic activity in the brain.

(1) It is possible to increase the concentration of endogenous CA at the synapse, with the aid of amphetamines and l-DOPA. In response to amphetamine, a number of synaptic vesicles containing DA and NA release their contents into the synaptic cleft, and in addition the re-uptake of DA and NA into the neuron is inhibited so that more CA become available at the receptors (Fuxe and Ungerstedt, 1970). In response to the CA precursor l-DOPA, the DA concentration in the brain shows a marked increase, while that of NA increases but little (Hornykiewicz, 1966). The predilection for DA formation is possibly based on the fact that the enzyme DA-β-hydroxylase is already saturated under normal conditions.

(2) CA receptors can also be stimulated directly: the DA receptor with apomorphine (Andén et al., 1967), and the NA receptor with clonidine (Andén et al., 1970b), and this occurs rather selectively.

(3) Finally, DA or NA can be introduced directly into the brain with a cannula, either into the ventricles or locally, e.g., into the caudate nucleus, which contains numerous DA-ergic nerve endings. In the former case, there is a risk that CA penetrate also to sites at which they do not normally occur (Cools, 1973).

Increased intracerebral CA-ergic activity, however it is produced, has its implications in terms of motor activity: locomotor activity increases and abnormal movements, e.g., stereotyped movements, occur. These are movements which as such are normal but become abnormal because they are long repeated without apparent reason.

These motor effects of CA-ergic hyperactivity are arrested by neuroleptics—a fact which is consistent with the CA receptor block. This should be of a competitive nature because the blocking effect of neuroleptics on the effect of DA locally applied in the caudate nucleus can be abolished by another local application of DA (Randrup and Munkvad, 1970; Cools, 1973).

Neuroleptics induce symptoms of CA-ergic hypoactivity. In man and animal alike, neuroleptics induce symptoms known to accompany diminished CA-ergic activity. In animals, a state of diminished CA-ergic activity can be induced by:

(1) Inhibition of the synthesis of DA and NA (with the aid of a tyrosine hydroxylase inhibitor) or of NA alone (with the aid of a DA-β-hydroxylase inhibitor) (Rech et al., 1966; Svensson and Waldeck, 1970).

(2) Prevention of CA storage in synaptic vesicles, with the aid of such agents as reserpine or oxypertine. The CA are degraded by MAO and their concentration in the brain decreases (Bein, 1956; Hassler et al., 1970).

(3) Causing lesions of CA sites of predilection, usually by electrocoagulation. In this way, one can destroy, say, the substantia nigra (Sourkes and Poirier, 1966). This is the location of the cell bodies of DA-ergic neurons which extend fibers to the corpus striatum via the internal capsule. This intervention eliminates about 75% of the cerebral DA. NA sites in the brain are less well-defined, but we nevertheless know some sites of predilection, e.g., the locus coeruleus.

CA-ergic systems can also be chemically eliminated, with the aid of 6-hydroxy-DOPA and 6 hydroxy-DA compounds which are selectively taken up into CA-ergic nerve endings and destroy them. 6-Hydroxy-DA affects both NA-ergic and DA-ergic nerve endings; 6-hydroxy-DOPA is more selective and has a predilection for NA-ergic nerve endings. Introduced into the ventricles, these compounds eliminate the entire central CA-ergic system; locally applied, they cause a local lesion (Thoenen and Tranzer, 1973; Richardson and Jacobowitz, 1973).

A consistent phenomenon after elimination of CA-ergic transmission, however this is effected, is diminution of locomotor activity. And this also applies to human individuals, as shown in Parkinson's disease. The most consistent morphological change in this disease is degeneration of melanin-bearing neurons in the pars compacta of the substantia nigra; the most consistent biochemical change is decreased DA and HVA concentrations in the corpus striatum. Moreover, there is a positive correlation between numerical cell loss in the substantia nigra and DA deficiency in the corpus striatum (Hornykiewicz, 1972). In Parkinson patients, diminished locomotor activity is an essential symptom; moreover, it is the symptom which usually responds best to the CA precursor l-DOPA.

In short, diminished motor activity is a characteristic feature of the CA deficiency syndrome as well as of the action profile of neuroleptics. This corroborates the conception of neuroleptics as CA antagonists.

Neuroleptics and cyclic AMP. There is a more direct indication that neuroleptics block CA receptors. This has recently been carefully established in particular for DA (review by Iversen, 1975). It is likely that cyclic nucleotides are involved in impulse transmission by means of DA and NA. It is assumed, for example, that cyclic adenosine monophosphate (cyclic AMP) is synthesized in the activation of postsynaptic CA receptors (Kebabian et al., 1972). In the brain, there are substantial regional differences in the synthesis of cyclic AMP in response to the various CA. These amines generally cause cyclic AMP accumulation only at sites at which they are active. The synthesis of cyclic AMP in response to DA is very markedly suppressed by drugs with a neuroleptic action, but not, or hardly at all, by compounds of related chemical structure but without neuroleptic activity (Clement-Cormier et al., 1974; Miller et al., 1974; Karobath and Leitich, 1974).

Other effects of neuroleptics on DA-ergic transmission. Seeman and Lee (1975) found indications that neuroleptics cause presynaptic block by inhibiting the release of DA from the synaptic vesicles after stimulation of the axon. Their view is based on *in vitro* work with rat brain slices which are electrically stimulated. The concentrations of neuroleptics required for 50% inhibition of the (^{3}H) DA release corresponded well with the mean daily dosages in clinical use; with clinically inactive isomers, amounts required were 20-1,000 times as large. For inhibition of the release of (^{3}H) acetylcholine and (^{3}H)GABA (γ aminobutyric acid), much larger doses of neuroleptics were required than for that of DA release. The phenomenon therefore seems to be fairly specific.

The increased DA turnover caused by neuroleptics is usually related to the increased firing rate in the presynaptic element, which should possibly be regarded as an attempt to break the postsynaptic receptor block. However, this cannot be the only explanation, for neuroleptics also increase DA turnover (if slightly less) when the axons of the DA neurons have been severed, thus eliminating an increased firing rate (Kehr, 1972). It is conceivable that the increase in turnover is in part secondary to inhibition of DA release, and one can imagine that this is the mechanism that persists after axon severance.

We are probably dealing with presynaptic as well as postsynaptic block, for presynaptic receptor block alone cannot explain that neuroleptics antagonize the effects of apomorphine—a compound which directly stimulates postsynaptic DA receptors. Be this as it may, receptor block and inhibition of release in any case both lead to diminished transmission in DA-ergic neurons.

Differential Influence of Neuroleptics on DA and NA Receptors

Ratio DA:NA receptor block. Do neuroleptics equally block DA and NA receptors, or do they differ in their affinity for these structures? The principal pertinent study available is that of Andén et al. (1970a). As a measure of DA receptor block, they accepted the following test arrangement. Rats with a unilaterally destroyed corpus striatum turn head and tail (a) away from the side of the lesion in response to DA-antagonistic compounds such as neuroleptics, and (b) to the side of the lesion in response to DA-agonistic compounds such as apomorphine. The latter effect is abolished by neuroleptics or, more generally, by compounds which block DA-ergic transmission. As a measure of DA receptor block, therefore, they accepted the extent to which a neuroleptic induces an asymmetrical posture and blocks apomorphine-induced turning in rats with a unilaterally destroyed corpus striatum.

The criterion accepted for NA receptor block was the following. After stimulation of the foot-sole of a spinal animal, the hindleg flexes—a reflex which is enhanced by l-DOPA. This potentiation by l-DOPA is antagonized by compounds which interfere with NA-ergic transmission.

The study of Andén et al. led to a division of neuroleptics into three groups.

(1) Compounds which exclusively block DA receptors and have no effect on NA receptors. This group includes compounds of the diphenylbutylpiperidine category, i.e., pimozide (Orap) and fluspirilene (Imap).

(2) Compounds which vigorously block DA receptors and only slightly block NA receptors. This group comprises representatives of the phenothiazines, e.g., perphenazine (Trilafon); the thioxanthenes, e.g., clopenthixol (Sordinol); and the butyrophenones. e.g., haloperidol (Haldol).

(3) Compounds whose DA receptor-blocking and NA receptor blocking potencies are of the same order. This group includes such compounds as the phenothiazine derivative chlorpromazine (Thorazine) and the thioxanthene derivative chlorprothixene (Truxal).

Neuroleptics increase the cerebral MHPG concentration in rats, but to a varying extent (Keller et al., 1973). This fact, too, indicates the probability of a widely diverse influence on NA turnover.

Biochemical vs. motor action profile of neuroleptics. The conventional classification of neuroleptics is based on chemical structure. This classification principal has been used in virtually all research into differences in effect between different neuroleptics. The above-discussed study by Andén et al. (1970a) has shown, however, that neuroleptics can also be classified in biochemical terms, and that this classification is quite at odds with that on the basis of chemical structure.

Is there a correlation between the biochemical classification of neuroleptics and their motor effects? We do not know. Research into this question in human subjects has so far been sporadic. The results obtained would seem to suggest such a correlation, but in this context a handful of data cannot be accepted as conclusive. In animal experiments, the question has likewise received scanty attention, and no sophisticated ethological behavior analyses have been made (O'Keeffe et al., 1970). On the other hand, a considerable body of research has been devoted to the preeminent question: Which role do DA and NA play in regulation of motor activity? Yet this question, too, has remained moot. The next section presents some notes on this research.

The Role of DA and NA in Regulation of Motor Activity

As pointed out, CA-ergic systems play a role in motor regulation. It can be maintained in general that their activation increases motor activity, and that their block reduces motor activity. The distribution of the roles of DA and NA in this respect, however, is still obscure. There are numerous fragmentary data, and there are two hypotheses which, in principle, seem acceptable. The first hypothesis postulates that stereotyped behavior is brought about by activation of DA receptors in the corpus striatum. The second assigns to the NA-ergic system a predominant role in regulation of the amount of (normal) locomotor activity. These hypotheses arise from work with amphetamine derivatives (Randrup and Munkvad, 1966; Randrup and Scheel-Krüger, 1966). These compounds modify behavior in test animals in two ways: they provoke stereotyped behavior (which differs in type from one species to the other), and they increase locomotor activity. Prior administration of a tyrosine hydroxylase inhibitor, producing reduction of the amounts of DA and NA available, causes *both* behavior changes to remain absent. Premedication with a DA-β-hydroxylase inhibitor, which inhibits only the NA production, causes amphetamine admini- stration to produce stereotyped behavior but not hypermotility.

In principle, these hypotheses are plausible. If DA-ergic neurons in the corpus striatum are stimulated by means other than amphetamine administration, e.g., by electrical stimulation or with apomorphine (a stimulator of DA recep- tors), then stereotyped behavior occurs as expected. It remains absent, however, after previous destruction of the corpus striatum (Cools, 1973).

The position of the hypothesis on the role of NA was strengthened when it was found that infusion of NA and 6-hydroxy-DOPA (destruction of NA-ergic nerve endings) into the lateral ventricles of the rat enhances and reduces motor activity, respectively (Segal and Mandell, 1970, Richardson and Jacobowitz, 1973).

NA plays a role in the "production" of motor activity, and DA in the com- pletion of an action. This conclusion seems justifiable. These roles, however, are not exclusive roles. After stimulation of NA-ergic neurons with small doses of morphine, for example, stereotyped behavior is also observed (Ayhan and Ran- drup, 1973), and the same is observed after activation of 5-HT neurons with fenfluramine (Southgate et al., 1971) or quipazine (Costall and Naylor, 1975). On the other hand, there are numerous data which indicate that, apart from NA, DA is also involved in locomotion. For example, a substance such as pimozide— a neuroleptic which selectively blocks DA receptors—can be used to block hyper- motility induced by amphetamines (Ayhan and Randrup, 1973).

It is for this reason that the hypotheses discussed were referred to merely as plausible "in principle."

Neuroleptics and CA Stores

Neuroleptics of the phenothiazine type (and the chemically related cate- gory of the thioxanthenes) and of the butyrophenone type (and the chemically related category of the butylphenylpiperidines) probably block central post- synaptic DA and NA receptors and, therefore, transmission in these systems. Apart from these, there are two other types of neuroleptics—reserpine (one of the Rauwolfia alkaloids) and oxypertine (an indole derivative)—which do not block receptors but exert an influence on the central CA metabolism. Reserpine inhibits uptake of CA into the synaptic vesicles, thus exposing them to MAO and reducing the amount of transmitter substance available for transmission. 5-HT storage is likewise inhibited, but the motor effects of reserpine are proba- bly based chiefly on the inhibited CA-ergic transmission (Carlsson et al., 1974).

Oxypertine likewise interferes with the intraneuronal storage of MA, but it is more selective than reserpine. Within a given dosage range, it shows a predilec- tion for the NA stores while exerting relatively little influence on DA and 5-HT storage (Hassler et al., 1970). It therefore suppresses NA-ergic transmission more or less selectively.

Conclusions

Neuroleptics of the chlorpromazine and the butyrophenone type (and the respective groups of the thioxanthenes and diphenylbutylpiperidines which are related in chemical structure) increase the intracerebral turnover of DA and NA—an effect which is probably secondary to (a) postsynaptic DA and NA receptor block, and (b) inhibition of DA release into the synaptic cleft after stimulation of the axon. The extent to which the various neuroleptics influence DA and NA receptors, respectively, differs widely. This biochemical differentiation does not coincide with the classification on the basis of chemical structure.

The remaining neuroleptics, i.e., reserpine (one of the Rauwolfia alkaloids) and oxypertine (an indole derivative), do not block CA receptors. They inhibit the uptake of CA into the stores, causing an increased degree of degradation by MAO and reducing the amount of transmitter substance available for release.

Thus all the known neuroleptics inhibit transmission in CA-ergic neuronal systems, be it via different mechanisms.

CHAPTER XVII

Neuroleptics and CA-ergic Transmission in the Brain
Clinical Studies

CSF Studies

An impression of the human central MA metabolism can be gained by measuring the concentration of MA metabolites in the CSF. So far as the metabolism of 5-HT and DA is concerned, the informative value of this determination is substantially enhanced when probenecid is administered in advance. Probenecid inhibits the transport of 5-HIAA and HVA from the CNS to the blood stream. This leads to accumulation of these acids at a rate which, at least during a number of hours, equals their rate of production (i.e., the rate of degradation of the respective mother amines, 5-HT and DA). The amounts of 5-HIAA and HVA which accumulate in the lumbar CSF during a given period following probenecid administration thus provide an index of the amounts of 5-HT and DA, respectively, which have been metabolized in the CNS during this period. The value and limitations of the probenecid technique have elsewhere been discussed in detail, as have its advantages over the method of measuring baseline metabolite concentrations (chapter X).

Neuroleptics of the DA receptor-blocking type. The baseline HVA concentration in the CSF is increased by neuroleptics of the phenothiazine and butyrophenone type (Persson and Roos, 1969; Chase et al., 1970; Sedvall et al., 1975); the same applies to the probenecid-induced HVA accumulation (Bowers, 1972,

1973, Chase, 1972; Van Praag and Korf, 1975). The last-mentioned phenomenon is dose-dependent, and more pronounced after haloperidol than after chlorpromazine (Van Praag and Korf, 1975). These findings warrant the hypothesis that the central DA turnover is increased by the neuroleptics mentioned. In animals, the increased DA turnover is secondary to postsynaptic DA receptor block. The influence which neuroleptics exert on the serum prolactin concentration suggests that the same applies to human subjects.

The 5-HIAA response after probenecid is not influenced by neuroleptics, which suggests that the central 5-HT metabolism remains intact (Van Praag and Korf, 1975). This is in agreement with corresponding findings in animals.

Transport of (free) MHPG is not inhibited by probenecid; in fact, there is no known effective inhibitor of this transport process. We must therefore rely on measuring baseline concentrations. Only chlorpromazine has been studied, and this does not influence the MHPG level (Van Praag and Korf, 1975). The conclusion that the NA turnover is not influenced, however, would be premature, for the baseline concentration of a MA metabolite is a much cruder and less reliable index of central MA metabolism than are post-probenecid values.

Neuroleptics of the MA store-depleting type. Of this group, only oxypertine has been studied in human individuals (Van Praag and Korf, 1975). It does not alter the probenecid-induced accumulation of 5-HIAA and HVA, which suggests that it exerts little influence on 5-HT and DA metabolism. However, the MHPG level shows a significant increase. This is consistent with the view that oxypertine depletes NA stores, causing the NA concentration to decrease and that of NA metabolites to increase. The human findings thus confirm expectations based on results of animal experiments.

Prolactin as a Marker of Central DA-ergic Activity

Numerous indications suggest that DA-ergic neurons of the tubero-infundibular system are involved in the synthesis and release of prolactin by the anterior pituitary (Meites et al., 1972; Schally et al., 1973). The neurons are believed to stimulate the release of the so-called prolactin-inhibiting factor (PIF) stored in the median eminence. Via the portal system of the hypophysis, PIF is transported to the anterior pituitary, where it inhibits the synthesis and/or release of prolactin, and thus causes the serum prolactin level to decrease. Since prolactin is exclusively produced in the anterior pituitary, its serum concentration can be regarded as an indirect measure of DA-ergic activity in the tubero-infundibular system. Without an attempt at comprehensiveness, I shall list the arguments on which this theory is based.

Drugs which reduce the amount of CA available at the postsynaptic receptors, such as the CA synthesis inhibitor α-MT, cause an increase in serum prolactin concentration in rats, and a decrease in prolactin in the anterior pituitary (Lu et al., 1970). Drugs which increase the central CA concentration, such as l-DOPA, reduce the serum prolactin concentration and increase that in the anterior pituitary. The former has been demonstrated in rats and human subjects, and the latter of course only in rats (Lu et al., 1970, 1971; Kleinberg et al., 1971). These studies do not permit a conclusion on the question whether DA or NA is responsible for this effect, but other observations have labeled DA as the most plausible candidate. For example, 10 minutes after introduction of a small amount of DA into the rat third ventricle, the serum prolactin level shows a sharp fall. NA and adrenaline do not produce this effect (Kamberi et al., 1971) Moreover, the decrease in serum prolactin concentration produced by l-DOPA is not blocked by compounds which inhibit conversion of DA to NA (Donoso et al., 1971). Finally, the fact that the effect of DA on prolactin release is produced not directly but through the intermediary PIF is concluded from the fact that the serum prolactin concentration does not decrease if DA is introduced directly into the portal circulation of the hypophysis or into the arterial blood supply (Kamberi et al., 1971).

Now, neuroleptics cause a rapid increase in serum prolactin concentration. The effect becomes manifest after a single dose, but also after chronic administration, and it has been observed in human and animal test subjects after neuroleptics of the receptor-blocking type (phenothiazines) as well as of the store-depleting type (reserpine) (Lu et al., 1970; Donoso et al., 1971; Kleinberg et al., 1971; Meltser et al., 1975). In human subjects, the prolactin concentration in lumbar CSF also increases (Sedvall et al., 1975). These findings can be regarded as an indirect indication that neuroleptics suppress the central activity in the tubero-infundibular DA system.

Other transmitter substances also influence the serum prolacting concentration. 5-HT is one. Introduced into the rat ventricle, it increases the serum prolactin concentration (Kamberi et al., 1971). The 5-HT precursor 5-HTP has a similar effect in rats (Lu and Meites, 1973). It has therefore been postulated by some investigators that 5-HT facilitates the release of a prolactin-releasing factor which counteracts the effect of PIF. Our observations, however, contradict the results of animal experiments. After infusion of 50 mg l-5-HTP in normal test subjects pre-medicated with a peripheral decarboxylase inhibitor, the serum prolactin concentration showed a marked, prolonged decrease. 5-HTP seemed to behave like a DA-agonist—an effect we have so far failed to explain (Van Praag, et al., 1976).

Cholinergic systems probably also play a role in the central regulation of prolactin secretion. Atropine, for example, blocks the midday peak in rat serum prolactin concentration as well as the preestrus peak (Libertum and McCann, 1973). It is unlikely, however, that it could be these effects that underlie the changes caused by neuroleptics in the serum prolactin concentration for they leave the central 5-HT metabolism intact. Should the anticholinergic properties which many neuroleptics possess be responsible for the prolactin effect, then a decrease rather than an increase in prolactin concentration would be expected.

Duration of the Effect of Neuroleptics on DA-ergic Transmission

The influence of neuroleptics on the probenecid-induced accumulation of HVA gradually diminishes in the course of a few weeks, and accumulation is normalized after 3 weeks (Post and Goodwin, 1975b) This does not mean that the block of the postsynaptic DA receptors has been abolished, for the prolactin effect persists undiminished (Van Praag, 1977a). The compensatory increase in DA turnover apparently subsides after a few weeks. This may explain the fact that the optimal therapeutic effect of neuroleptic medication is not obtained immediately, but only after a few weeks. As long as DA production is increased, after all, it is possible that the receptor block is partly or completely overcome. For this block is of a competitive nature, which means that it can be overcome by a surplus of substrate. Conceivably, inhibition of transmission attains its maximum only when receptor block is associated with normalized DA synthesis. An argument in favor of this theory is the fact that the therapeutic effect of neuroleptics is potentiated when they are combined with α-methyl-p-tyrosine—an inhibitor of DA and NA synthesis (Carlsson et al., 1972).

Conclusions

Neuroleptics of the receptor-blocking type increase both the baseline HVA accumulation and the post-probenecid HVA accumulation in human CSF. This would seem to suggest that, as in test animals, they increase the central DA turnover. In the human organism, too, this phenomenon is probably secondary to inhibition of transmission in the DA-ergic system. This is concluded from the observation that neuroleptics increase the serum prolactin concentration—an effect indicative of reduced DA-ergic activity in the tubero-infundibular system. Oxypertine has no demonstrable effect on human central 5-HT and DA metabolism, but increases the MHPG concentration in the CSF. This finding is consistent with depletion of central NA stores.

The human findings correspond well with those obtained in animal experiments, and the hypothesis that neuroleptics block CA-ergic transmission also in the human brain therefore seems justifiable.

CHAPTER XVIII

Reduced CA-ergic Transmission and Clinical Effects of Neuroleptics

Clinical Effects of Neuroleptics

Therapeutic effects. Neuroleptics have a dual effect in psychoses: a sedative and an antipsychotic effect. The sedative effect is generally strong, whereas the hypnotic effect is modest in comparison with that of the classical sedatives such as barbiturates and bromides. The so-called antipsychotic effect is a very complex one. In some cases, the term "antipsychotic" can be translated into "de-emotionalizing." The affective response to various psychotic symptoms such as delusions and hallucinations diminishes, although the symptoms per se do not (entirely) disappear. The patient's delusions and hallucinations continue, but he is less affected by them; he can remain aloof from them; he is less agonized by them. In other cases, there may be literally an antipsychotic effect: the psychotic symptoms disappear and the patient returns to the premorbid level. This can be a very rapid return, within a day or a few days, or the effect becomes manifest more gradually, in the course of a few weeks. This cannot be described as a real latent phase, of the kind seen in antidepressant medication. If the medication "takes," then some effect is apparent within the first few days, and gradually increases.

We have as yet no criteria for a prediction whether a de-emotionalizing or a real anti-psychotic effect is likely to occur. In some cases, neuroleptic medication, even in large doses, fails; in such cases, the psychotic symptoms may in fact show exacerbation, possible as a result of intoxication. Such responses, too, cannot be predicted.

Some patients show a much better response to one neuroleptic than to the other, even when they are classified in the same psychopathological category. In any case, the choice of a neuroleptic is a rather arbitrary one. Actually, the only guideline is the extent to which the drug suppresses motor activity and initiative. Markedly sedative neuroleptics (such as the phenothiazines with a terminal aminodimethyl group in the side chain) should be prescribed in the case of severe motor unrest. Neuroleptics which are less sedative or which may even be slightly activating (such as the phenothiazines with a piperazine ring in the side chain) are indicated in the treatment of chiefly inert patients.

Let me repeat: we do not know the causes of uncertainties in neuroleptic medication. There are several possibilities: the "climate" in the ward in which the patient is being treated; the etiology of the psychosis; its pathogenesis. By "pathogenesis" I mean the cerebral substrate which enables psychotic symptoms to occur; by "etiology" I mean the complex of factors (psychological, social, genetic and acquired somatic) which have contributed to the formation of the cerebral substrate (Van Praag and Leijnse, 1963, 1965). As regards the factor pathogenesis: psychotic syndromes vary widely in symptomatology. In other words, they can involve a disturbance of widely diverse psychological functions. This warrants the suspicion that the underlying cerebral substrate, too, must be variable. In principle, the concept of a biochemical typology of psychoses strikes me as no less real than that of such a typology of depressions (Van Praag and Korf, 1971; Van Praag 1974b).

Motor side effects. All neuroleptics, regardless of their chemical structure, can provoke extrapyramidal symptoms. These are interpreted as untoward side effects, and controlled by means of the conventional anticholinergic anti-Parkinson drugs. It is generally assumed that these drugs do not detract from the therapeutic efficacy of neuroleptics. Perhaps this view needs revision, for it was recently reported by Singh and Kay (1975) that the efficacy of neuroleptic medication certainly diminishes in response to benztropine.

All neuroleptics are capable of provoking hypokinetic-rigid symptoms. In this respect, there are no consistent differences between the different chemical groups (Van Praag and Korf, 1975). The occurrence of hyperkinetic and dyskinetic symptoms is more dependent on given chemical categories, specifically the piperazine derivatives of the phenothiazine series, and the butyrophenones. We do not know the reason for these group differences.

Animal Experiments

DA receptor block and therapeutic effects of neuroleptics. Animal experiments have yielded some indications that inhibition of transmission in CA-ergic systems does indeed correspond with the therapeutic efficacy of neuro-leptics. To begin with, the central DA turnover is not increased by phenothiazine derivatives without neuroleptic potency, whereas neuroleptic phenothiazines do cause such an increase (Carlsson and Lindqvist, 1963; Andén et al., 1964; Nybäck et al., 1968; Matthysse, 1973). The same applies to their influence on the DA-sensitive adenylcyclase in the caudate nucleus: this enzyme is inhibited by neuroleptic phenothiazines, but not by phenothiazines without neuroleptic activity (Kebabian et al., 1972). Of all the biological effects of neuroleptics studied, moreover, this effect occurs at by far the lowest concentrations (Iver-sen, 1975). There are no other known biochemical and biophysical effects of phenothiazine derivatives which differentiate anti-psychotic from non-anti-psychotic compounds (Matthysse, 1973). This is a strong indication that an influence on the DA system is a prerequisite for production of the therapeutic effect.

Moreover, there is a fair degree of agreement between the ranking of neuro-leptic phenothiazines according to DA receptor-blocking ability and a ranking of these compounds according to average clinical dosage used. The DA receptor-blocking ability was assessed on the basis of the increase in central HVA concen-tration (Nybäck et al., 1970a; Rollema and Westerink, 1976), on the basis of their inhibitory effect on DA activation of DA-sensitive adenylcyclase (Clement-Cormier et al., 1974, Horn et al., 1974; Miller et al., 1974), and by direct analysis of the DA-receptor (Seeman et al., 1975).

Of course, these data are not conclusive of a causal relation. To validate such a conclusion, two series of data would have to be available: (a) dose-effect curves of a series of neuroleptics in terms of their stimulating effect on the central DA turnover, (b) dose-effect curves of the same series of neuroleptics in terms of their anti-psychotic effect, the therapeutic optimum (i.e., the dose with the highest therapeutic efficacy) having been established for each agent. A fair degree of correspondence between these two rankings would indicate the likelihood of a causal relation between DA effect and therapeutic effect of neuroleptics. For such an analysis, however, no adequate data are available: those referred to sub (1) are available for a number of neuroleptics, but those referred to sub (2) are lacking. Assuredly, we know that, on a milligram basis, one neuroleptic is more potent than the other. For example, the average daily dose prescribed is 3.5 mg/kg for chlorpromazine and 0.05 mg/kg for haloperidol (Davis, 1974). But this does not mean that the therapeutic efficacy of halo-peridol exceeds that of chlorpromazine. For a ranking of neuroleptics according to therapeutic efficacy, the necessary data are lacking.

Neuroleptics, both of the receptor-blocking and of the store-depleting type, induce a characteristic paucity of movement (catalepsy) in test animals. This syndrome is antagonized by drugs which facilitate CA-ergic transmission, e.g., amphetamines; on the other hand, the hyperactivity induced by the latter com-pounds is arrested by neuroleptics (Gordon, 1967). Consequently, it is plausible that at least the motor-sedative effect of neuroleptics is related to their depressant effect on CA-ergic transmission.

Little is known about the significance of inhibition of NA-ergic transmission for the therapeutic effect of neuroleptics. An increased NA turnover is observed, not only after neuroleptic phenothiazines but also after promethazine: a phenothiazine without neuroleptic properties (Nybäck et al., 1968). But this does not mean that this effect is of little significance for the neuroleptic effect, as Matthysse (1973) concluded. Though promethazine may not be a neuroleptic, it does have a sedative effect. It is therefore quite possible that the NA receptor block is related to the sedative effect of neuroleptics (Van Praag et al., 1977).

DA receptor block and motor side effects of neuroleptics. All neuroleptics can cause extrapyramidal side effects. DA functions as a neurotransmitter in the extrapyramidal system. The effect of neuroleptics on DA turnover may therefore be correlated with the motor side effects, not with the therapeutic effects. However, this is unlikely. There are two known neuroleptics which, used in the conventional doses in human subjects, cause extrapyramidal symptoms only sporadically and, if so, rarely in any serious degree. These agents are thioridazine (Mellaril) and clozapine (Leponex). Nevertheless, both compounds increase the HVA concentration: in test animals, in the caudate nucleus (Roos, 1965; O'Keeffe et al., 1970; Andén and Stock, 1973; Westerink and Korf, 1975a), and in human subjects, in the lumbar CSF: the baseline concentration (Gerbode and Bowers, 1968; Ackenheil et al., 1974) as well as the post-probenecid concentration (Van Praag, et al., 1977). All these findings are suggestive of an increased DA turnover. The findings of Gerlach et al. (1975), who observed a decrease in baseline HVA concentration in response to clozapine in 6 out of 8 chronic schizophrenics studied, stand out as being apparently inexplicable.

However, there is a difference between thioridazine and the standard phenothiazine-type neuroleptic chlorpromazine: the HVA concentration in the feline caudate nucleus is increased both after a single dose and after chronic administration of chlorpromazine, whereas this effect soon disappears after chronic thioridazine administration (Laverty and Sharman, 1965). Moreover, in the acute experiment, the HVA effect of chlorpromazine is likewise substantially more intensive than that of thioridazine and clozapine (Matthysse, 1973; Crow and Gillbe, 1973; Westerink and Korf, 1975a). This could indicate that the increase in DA turnover is after all chiefly related to the motor symptoms.

Yet another explanation is possible. The bulk of the central DA is contained in the nigrostriatal system (70-75%), while only 20-25% is localized in the mesolimbic and mesocortical systems. If the neuroleptics in question would exert a less marked influence on the former system than on the latter, then the small overall effect on the central HVA concentration would be explained. Such regional differences have in fact been described for clozapine (Andén and Stock, 1973; Bowers and Rozitis, 1974), and have supported the hypothesis that inhibition of DA-ergic transmission in the nigrostriatal system underlies the motor side effects of neuroleptics, while inhibition of DA-ergic transmission in the mesolimbic and mesocortical systems is held responsible for their therapeutic effects. However, these observations have not yet been corroborated. With a series of neuroleptics, including thioridazine and clozapine, the percental increase in HVA in both DA systems was virtually the same per neuroleptic (Westerink and Korf, 1975a; Wiesel and Sedvall, 1975). In other words, the HVA effect of clozapine and thioridazine was small in both DA systems. Assuming these neuroleptics to be fully valid therapeutically, if not very active extrapyramidally, this is an argument against a correlation between DA effect and therapeutic effect. According to our data, however, the therapeutic validity is dubious. In a comparative study of clozapine and perphenazine, the former compound was found to be a strong sedative but inferior to perphenazine in anti-psychotic effect (Van Praag et al., 1977). According to Cole et al. (1964), thioridazine is equivalent to chlorpromazine. According to our data, however, thioridazine is inferior as a standard neuroleptic.

There is *no* discrepancy, therefore, between our findings and the hypothesis that DA receptor-blocking potency and therapeutic (i.e., anti-psychotic) effect of neuroleptics are related. Moreover, they warrant the expectation that a neuroleptic with limited extrapyramidal side effects will also have a limited anti-psychotic effect. Continued comparative studies of the therapeutic "weight" of clozapine and thioridazine will have to show whether these statements are sound.

Clinical Studies

DA receptor block and therapeutic effects of neuroleptics. There are two direct indications that the two phenomena are related. The first is that promethazine (Fenergan), a phenothiazine derivative without anti-psychotic properties, does not increase the HVA response in the CSF to a probenecid load (Van Praag, unpublished observations). As is the case in animals, therefore, the stimulant effect on the DA turnover seems to be a property of phenothiazines with an anti-psychotic effect.

The second argument arises from a study of 32 patients suffering from acute schizophrenic psychoses of varying symptomatology and etiology, who were treated with chlorpromazine (Thorazine), haloperidol (Haldol) or perphenazine (Trilafon), and in whom the therapeutic effect of this medication was compared with the degree of increase in the HVA response to probenecid during medication (Van Praag and Korf, 1975, 1976a). The neuroleptic dosage was daily adjusted to requirements by a psychiatrist not involved in behavior evaluation. Neither patients nor raters were aware of the type of neuroleptic used, its dosage, and the results of the probenecid test. A positive correlation was found between the degree of improvement during the second week of medication and the percental increase in HVA response in relation to the pre-therapeutic value. In other words, clinical improvement was as much more pronounced as the increase in HVA response was more marked. This relation proved not to be determined only by the sedative action component of the neuroleptics (i.e., their tranquilizing effect), for it persisted when, instead of the total scores, only the improvement in the two prototypical psychotic symptoms—delusion and hallucination—was taken into account. According to Sedvall et al. (1975) there is also a positive correlation between the increase in the baseline HVA concentration in the CSF, and the therapeutic effect of neuroleptics.

These observations strongly suggest that increased DA turnover (read: reduced DA-ergic transmission) and therapeutic efficacy of neuroleptics are related.

After longer periods of treatment (3-6 weeks) with phenothiazines, the effect on DA turnover subsides (Post and Goodwin, 1975). The serum prolactine level, however, remains high (Van Praag 1977a); apparently, the DA-receptor block continues to exist. One could assume that only after termination of the compensating increase in DA synthesis the pre- and postsynaptic DA functions would be optimally reduced. This could explain the fact that maximum clinical effectiviness of neuroleptic treatment usually is obtained not immediately but after longer administration.

DA receptor block and motor (side) effects of neuroleptics. In view of the findings obtained in Parkinson's disease (review by Hornykiewicz, 1972), it seems plausible for the time being that the extrapyramidal symptoms which neuroleptics can provoke are related to (nigrostriatal) DA receptor block.

The most constant biochemical finding in Parkinson's disease is the decreased concentration of DA, HVA and DOPA decarboxylase in the nigrostriatal complex. This decrease is considered to be based on degeneration of melanin-bearing neurons in the substantia nigra, being the most consistent morphological change in this syndrome. The correlation between the two phenomena is the more plausible because the degree of cell loss in the substantia nigra correlates with the degree of striatal DA deficiency. *In vivo,* the DA deficiency manifests itself in a reduction of the HVA response in the lumbar CSF to probenecid, which has been observed in these patients (Olson and Roos, 1968). The pathogenetic importance of the DA deficiency is apparent from (a) the therapeutic efficacy of l-DOPA, a DA precursor which is converted largely to DA but only in small part to NA (Cotzias et al., 1967; (b) the fact that the DA deficiency in hemi-Parkinson patients is localized chiefly in the corpus striatum contralateral to the side of the symptoms (Barolin et al., 1964; (c) the negative correlation found to exist between the pre-therapeutic HVA response to probenecid and the therapeutic effect after six months' medication with l-DOPA (Lakke et al., 1972; Korf et al., 1974). The effect was as much more marked as the pre-therapeutic HVA response was smaller.

However, the correlation between DA and Parkinson's disease has only partial relevance to the crucial question: whether the extrapyramidal symptoms caused by neuroleptics are based on a functional DA deficiency. For it is to be noted that (a) neuroleptics provoke not only hypokinetic but also hyperkinetic extrapyramidal symptoms, and there are no indications that extrapyramidal hyperkinesias of non-neuroleptic origin are based on a central DA deficiency; (b) l-DOPA is ineffective against non-neuroleptic extrapyramidal hyperkinesias; (c) a given dose of a given neuroleptic provokes extrapyramidal symptoms in some, but not in other patients, although the therapeutic effect observed warrants the conclusion that a sufficient concentration has reached the brain. Evidently, there is an individually very variable susceptibility to motor side effects.

Be this as it may, a correlation between at least neuroleptic parkinsonism and DA deficiency would be plausible if l-DOPA were effective in these cases. Unfortunately, there are few pertinent data. L-DOPA has been little used for this indication, because it can aggravate the psychotic symptoms. The available data suggest that l-DOPA counteracts hypokinetic-rigid symptoms provoked by neuroleptics (Bruno et al., 1966).

Another question of crucial importance in this context is whether a correlation exists between CSF HVA (as a measure of central DA metabolism) and neuroleptic Parkinsonism. Chase et al. (1970) studied a group of chronic psychiatric patients treated with various types of neuroleptics for at least a year. In patients without extrapyramidal symptoms, the baseline HVA concentration in the lumbar CSF was markedly increased. This level was much lower in patients with parkinsonoid and dyskinetic symptoms (although still above normal). Chase suggested that the CSF HVA decreases because DA-ergic neurons have been damaged by neuroleptics. Yet these observations provide no answer to the question as to whether, in the *actue* case, the increased DA turnover (interpreted as index of inhibited transmission) correlates with a possible occurrence of hypokinetic-rigid symptoms.

It seems likely that it does. During the second week of medication with neuroleptics of varying types, the HVA response to probenecid was found to be more marked in patients who had developed hypokinetic-rigid symptoms than in those without motor pathology. In addition, it was established that, in patients who developed hypokinetic-rigid symptoms in the course of medication, the pre-therapeutic HVA response to probenecid had been less marked than that in patients without these symptoms (Van Praag and Korf, 1975, 1976a). The interpretation of the latter finding was that the risk of neuroleptic Parkinsonism increases by as much as the pre-therapeutic DA turnover is lower. This was viewed as an explanation of the individual variations in susceptibility to the motor side effects of neuroleptics.

In view of all this, it is plausible that the occurrence of hypokinetic-rigid symptoms during neuroleptic medication is related to changes in the DA metabolism. Many neuroleptics are not only DA antagonists but, in varying degrees, also acetylcholine antagonists. Snyder et al. (1974) assumed that it is the anticholinergic potency that determines whether a neuroleptic provokes marked or less marked extrapyramidal symptoms in patients. The more marked the anticholinergic effect, the less marked the extrapyramidal symptoms. The above data are at odds with this hypothesis. They suggest that the degree of DA antagonism is certainly a determinant factor in the occurrence or nonoccurrence of extrapyramidal (or at least hypokinetic-rigid) symptoms.

The hyperkinesias and dyskinesias induced by neuroleptics have not been studied in this context. Chase (1972) assumed that the occurrence of hypokinetic or hyper(dys)kinetic symptoms is determined by the equilibrium between receptor block and increased DA turnover. If the former is predominant, he believed, hypokinetic-rigid symptoms occur. If the increase in DA turnover is so marked as to break through the receptor block—if it overshoots its mark, so to speak—then hyper(dys)kinetic symptoms should occur. This hypothesis could be tested in humans. After all, we do have a (rather crude) yardstick of central DA turnover, as well as an index of postsynaptic DA receptor activity (serum prolactine). This has yet to be done.

Conclusions

A ranking of neuroleptics according to their ability to inhibit DA-ergic transmission roughly corresponds with their ranking according to average clinical doses used. This indicates the probability of a correlation between DA antagonism and therapeutic effect, but does not by any means prove this correlation. In animal experiments, this proof cannot be obtained as long as there are no data which warrant a ranking of neuroleptics according to therapeutic potency. However, clinical studies have shown that such a correlation is very plausible. The increase which neuroleptics induce in the CSF HVA concentration (both in the baseline and in the post-probenecid concentration) proved to show a positive correlation with the therapeutic efficacy of the medication, both in sedative and in anti-psychotic terms.

With regard to the question whether extrapyramidal effects of neuroleptics are determined by central DA deficiency, clinical studies have likewise been more elucidative than animal experiments (neuroleptics induce no classical extrapyramidal symptoms in animals). The occurrence of hypokinetic-rigid symptoms proved to show a positive correlation with the increase in HVA accumulation after probenecid. In addition, it was found that the risk of these side effects is as much higher as the pre-therapeutic HVA response is lower. Neuroleptic hyperkinesias and dyskinesias have not been studied in this context.

All these data warrant the conclusion that changes in central DA metabolism play a role in the production of the clinical (side) effects of neuroleptics.

CHAPTER XIX

The CA Metabolism in Patients with Schizophrenic Psychoses

In the approach to the question whether the CA metabolism is disturbed in schizophrenic patients, three research strategies were applied: enzyme studies, studies of CA metabolites in the CSF, and serum prolactin determinations. The results are presented in the following sections.

Monoamine Oxidase (MAO)

Functional MAO deficiency. A few years ago, Murphy and Wyatt (1972) studied a group of patients with schizophrenia, mostly of the chronic type, and found the MAO activity in the blood platelets decreased by an average of 50%. This phenomenon was not caused by psychotropic drugs, and the MAO activity per individual was fairly stable in time, and independent of therapeutic results. Moreover, in monozygotic twins of whom one was schizophrenic and the other not, the MAO activity in the platelets was found to be closely correlated (Wyatt et al., 1973). Decreased MAO activity, therefore, is not associated with schizophrenia as such but might be a genetic marker which determines the "vulnerability" to schizophrenic psychoses. Should the intracerebral MAO activity be decreased also, then this would be a factor predisposing to DA-ergic hyperactivity.

In a reduplication study, Carpenter et al. (1975) established that the phen-

212   SCHIZOPHRENIC PSYCHOSES

omenon occurs exclusively in chronic types of schizophrenia. This observation was confirmed in several independent studies (Meltzer and Stahl, 1974; Nies et al., 1974; Zeller et al., 1975). Zeller et al. (1975) maintained that there was no decrease in the concentration of the enzyme but an abnormality in its structure. According to Garelis et al. (1975), the serum 5-HT concentration is increased in chronic schizophrenic patients, as could be expected in view of a decreased MAO activity.

Counterarguments. Shaskan and Berkes (1975) have so far been the only dissenters in this context. They did find low MAO values in schizophrenic patients, but the same phenomenon was found in chronic alcoholics and normal volunteers. The cause of this discrepancy is obscure. Differences in technique of determination may have played a role, but also such intervening variables as iron deficiency —a factor which reduces MAO activity (Youdim et al., 1975); and finally it is conceivable that low MAO activity correlates, not with a given psychiatric syndrome, but with a particular (pathological or nonpathological) psychological phenomenon. The notion that a given biological variable must be nosologically or at least syndromally specific to be of significance, is persistent in biological psychiatry but, in my view, misleading. The concept of symptomatological specificity is perfectly justifiable, as I have explained elsewhere (Van Praag et al., 1975).

In any case, post-mortem examination of schizophrenic brains has so far failed to reveal MAO defects, either in absolute amount or in the pattern of the various enzyme types (Schwartz et al., 1974).

Dopamine-β-Hydroxylase (DBH)

DBH deficiency. In 1973, Wise and Stein published the startling observation that, in all cerebral structures examined in chronic schizophrenics, and specifically in diencephalon and hippocampus, DBH activity was reduced by about 50% vs. control values. DBH is an enzyme required to convert DA to NA, and a deficiency in this enzyme reduces the production of NA in NA-ergic neurons and increases that of DA. The resulting excess DA in NA-ergic neurons could start to function as a false transmitter and contribute to a decrease in neuronal activity in the NA-ergic system. It is also possible that some of this excess DA "flows over" and is taken up into DA-ergic neurons. This could explain the hyperactivity of the DA-ergic system which the DA hypothesis postulates in schizophrenia.

Theory of Stein and Wise. Stein and Wise (1971) considered the crucial point to be, not the excess of DA, but the deficiency in NA. The latter is an important transmitter in the so-called "pleasure" or "reward" systems. These are believed to generate the sense of gratification which follows certain activities, and which contributes significantly to their continuation or repetition. It was via hypofunction of these systems that Stein and Wise wanted to explain a number of essential symptoms of schizophrenia. The DBH deficiency observed was believed to be secondary to the production of a toxic CA metabolite, which is taken up into NA-ergic neurons and damages them. In this respect, they considered such substances as 6-hydroxy-DA, a synthetic DA derivative which has this property but whose *in vivo* production has never been demonstrated.

A more fundamental objection to this theory is the lack of any argument in support of the postulate of central NA deficiency in schizophrenic individuals. The MHPG concentration in the CSF is normal (Van Praag and Korf, 1975). Secondly, neuroleptics reduce not only DA-ergic but also NA-ergic transmission. The clinical significance of the last-mentioned phenomenon is obscure, but, if NA deficiency is a pathogenetic factor in schizophrenia, neuroleptics could be expected to produce an anti-therapeutic rather than a therapeutic effect. Evidently they do not. Thirdly, Olsson (1974) demonstrated by means of a histochemical technique that NA fluorescence in the schizophrenic brain does not differ from that in normal controls. Finally, this theory is not consistent with the results obtained with inhibitors of CA synthesis (α-MT) and DA agonists such as l-DOPA and amphetamines, which certainly potentiate NA as well as DA (Chapter XX).

Counterarguments. The only reduplication study so far made (Wyatt et al., 1975) failed to corroborate the findings of Wise and Stein. The DBH values were indeed decreased in schizophrenics (average 77-89% of control values), but the differences were not significant. Moreover, a negative correlation was found with the duration of the interval between death and post-mortem examination, and with the level of neuroleptic dosage. In other words, the DBH concentration was as much lower as the interval had been longer and the amount of neuroleptics used larger.

In their reply, Wise and Stein (1975) analyzed the statistical technique used and managed after all to extract some significance from the data reported by their opponents—which does not alter the fact that a 20% reduction of activity, even if real, would probably have no functional significance. They also discussed the possibility of a post-mortem artifact—the perennial flaw in this type of research. Unfortunately, they ignored the methodological problems of DBH determination (Laduron, 1975), to focus instead on the factor diagnosis. Leaving aside Wyatt's patients diagnosed as paranoid schizophrenics, and accounting only for the results obtained in what they diagnosed as chronic undifferentiated form of schizophrenia, they found that the DBH level was about 60% of control values, and approximated their own results. The significance of this last point of view is evident. It cannot be overemphasized that, in psychopathological terms, schizophrenia is the lid on a vessel full of widely diverse psychoses. The expectation that their pathogenesis is a uniform one strikes me as illusory, as does the hope of isolating biological factors characteristic of "schizophrenia."

The DBH controversy continues unsolved. In view of the importance of this question, further investigations are eagerly awaited.

Peripheral DBH activity. The available data on peripheral DBH activity do not point in the direction of a deficiency. Plasma DBH activity is normal in schizophrenic patients (Dunner et al., 1973, Shopsin et al., 1972). In monozygotic twins discordant for schizophrenia, the same has been found, along with a close correlation of their individual DBH values (Lamprecht et al., 1973).

Rosenblatt et al. (1973) studied the CA metabolism in sympathetic neurons of salivary glands, which contain an abundance of them. They administered ^{3}H-labeled DA to a number of psychiatric patients and normal test subjects, and measured the CA metabolism in the saliva. As index of DBH activity they used the percentage of measured radioactivity from NA metabolites. In patients with acute schizophrenia, they found a higher DBH activity than in depressive patients and normal test subjects. These observation as such do not necessarily contradict those of Wise and Stein. The latter's observations pertained to chronic patients, vs. acute patients in the Rosenblatt study. It is conceivable that DBH activity is increased in acute cases but extinguished in chronic cases. However, Rosenblatt et al. found increased DBH activity also in manic patients. This means that the phenomenon is not nosologically or syndromally specific, and may in fact be linked to the factor hyperactivity or to the degree of anxiety.

Methylene-THF Reductase

Freeman et al. (1975) described two sisters, aged 15 and 17, and a 16-year-old boy in another family, with an unusual form of homocysteinuria. Usually, accumulation of homocysteine in the organism and its increased renal excretion results from a deficiency in cystathionine synthase (or synthetase): the enzyme which catalyzes conversion of homocysteine to cystathionine (Fig. 1). In these three patients, however, there was a deficiency in a different enzyme: 5, 10-methylene-tetrahydrofolate reductase (methylene-THF-reductase), which regulates conversion of 5,10-methylene-THF to 5-methyl-THF. The latter compound is involved as methyl donor in the conversion of homocysteine to methionine. Deficiency in methylene-THF reductase leads to deficiency in 5-methyl-THF, thus to reduced conversion of homocysteine to methionine, and thus to homocysteinuria (Fig. 1). Administration of large amounts of folic acid, precursor of the THF derivatives, reduced the homocysteine excretion. Apparently, it increased the "flow" through the partly deficient reductase system, thus increasing the production of 5-methyl-THF.

The elder of the two sisters was suffering from a psychiatric syndrome with delusions, hallucinations, catatonic and autistic symptoms, which had been diagnosed as schizophrenia by several independent psychiatrists. She responded very favorably to folic acid, and relapsed when this agent was discontinued. Freeman

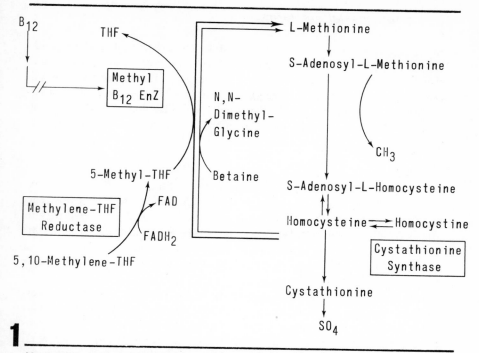

Metabolic pathways of homocysteine-methionine metabolism. Critical enzymes are enclosed in boxes (from Freeman et al., 1975).

et al. (1975), coined the term "folate-sensitive schizophrenia" (although they, placed it in quotation marks).

A bridge can be constructed between DA metabolism and methylene-THF reductase deficiency, for 5-methyl-THF could function also as methyl donor in the N- and O-methylation of IA and CA in the brain (Leysen and Laduron, 1974; Laduron et al., 1974a; Banejee and Snyder, 1973). A deficiently working reductase system, resulting in a deficiency in unstable methyl groups, could delay the degradation of MA, including that of DA, and increase the concentration available for transmission. In this suggestion, the core of the DA hypothesis is brought to view. Recently, however, this theory has been unsettled by a report by Meller et al. (1975), stating that 5-methyl-THF is not the favorite methyl donor in DA degradation. Levi and Waxman (1975) likewise postulated a deficiency in unstable methyl groups in schizophrenia. They held a deficiency in the classical methyl donor S-adenosylmethionine responsible, and contended that this deficiency in turn would result from defective functioning of the enzyme methionine adenosyltransferase. However, no facts indicative of such an enzyme deficiency have been presented.

The observation reported by Freeman et al. may be very interesting, but a deficient methylene-THF reductase system fails to afford an explanation for a substantial proportion of patients diagnosed as schizophrenic. In a series of 300 schizophrenic patients, not a single instance of homocysteinuria was found (Gershon and Shader, 1969). However, this observation is nevertheless of considerable theoretical significance if we assume that biochemical mechanisms of different types can ultimately lead to the same type of disorder in the CA metabolism, and that hypoactivity of the above-mentioned reductase system is one of them.

CSF Studies

Neuroleptics reduce: (a) agitation and psychotic disorders in thinking and experiencing; (b) neuronal activity in central DA-ergic (and NA-ergic) systems. This poses the question whether (some) psychoses perhaps involve hyperactivity in (certain) DA-ergic (and/or NA-ergic) systems. The suggestion is not an inevitable consequence of the factual observation. In principle, one can control a symptom and leave the disease process unaffected. An example is the use of antipyretics to control fever.

An increased baseline HVA concentration in the CSF is regarded as an indicator of an increased DA turnover in the CNS. This could be a result of increased DA-ergic activity. Several authors, however, have reported normal HVA concentrations in the CSF (Bowers et al., 1969; Persson and Roos, 1969). Rimon et al. (1971) reported increased values in paranoid types of schizophrenia, but this observation was not confirmed by Bowers (1973).

In psychotic patients without first-rank Schneider symptoms, post-probenecid HVA accumulation was found to be slightly more marked than in patients with these classical schizophrenic symptoms and in controls (Bowers, 1973). Post et al. (1975) found normal HVA accumulation in patients with acute schizophrenia. Yet the DA metabolism had not remained entirely unaffected, for after recovery the accumulation was significantly lower than before. Van Praag and Korf (1975) studied a group of patients with psychoses of mixed etiology but always with manifest delusions and/or hallucinations. The HVA accumulation was slightly, but not significantly, increased. After classification of the patients according to the degree of motor hyperactivity, HVA accumulation in agitated patients proved to be significantly higher than that in patients without this symptom. An increase in DA turnover, it was concluded, can occur in psychoses but is dependent on the factor motor activity rather than on "true" psychotic symptoms such as delusion and hallucination. Bowers found decreased post-probenecid values in lumbar CSF in one (Bowers, 1974), and increased values in another (Bowers, in press) of two recent studies in schizophrenic patients. This variability of Bowers' results also suggests that the disturbances in DA metabolism correlate less with the syndrome or disease entity schizophrenia than with a particular component (or with a particular subgroup still to be identified).

The MHPG concentration in the CSF (indicator of central NA metabolism) was normal (Post et al., 1975; Van Praag and Korf, 1975). The same applied to the serum prolactin level (Meltzer et al., 1975), and this parameter is regarded as an index of neuronal activity in the tubero-infundibular DA system. Finally, in a group of chronic schizophrenic patients, the activity of DA-sensitive adenyl cyclase in the caudate nucleus proved to be within normal limits (Carenzi et al., 1975).

There is no convincing evidence, therefore, that DA-ergic (or NA-ergic) hyperactivity plays a conclusive role in the pathogenesis of schizophrenia or other types of psychosis. However, it is to be borne in mind that HVA in the CSF originates largely from the nigrostriatal DA system, while prolactin release is controlled by the tubero-infundibular DA system. These measurements therefore give no information on local changes in the mesolimbic DA system or in the DA-ergic neurons which project on the cortex.

Conclusions

The view that reduction of central DA-ergic activity can result in a therapeutic effect on psychoses is a fairly sound one. But there is no convincing evidence to support its complement: the hypothesis that DA-ergic hyperactivity plays a role in the pathogenesis of schizophrenia or, more widely, psychotic symptoms. The enzymes studied in this context have yielded some intriguing results: reduced MAO and DBH activity in the blood platelets and in the brain, respectively, of schizophrenic patients (both controversial observations). And in the only pertinent study so far published, intracerebral MAO activity was found to be normal. One patient has been described with methylene-THF reductase deficiency, who developed a psychosis diagnosed as schizophrenia which responded favorably to large doses of folic acid—a medication which more or less compensates the enzyme deficiency. Interesting as this observation as such may be, it is fairly certain that this mechanism plays no role in the majority of schizophrenic patients.

CSF studies and prolactin determinations, too, have failed to yield indications that hyperactivity in DA-ergic and/or NA-ergic systems plays a decisive role in the pathogenesis of schizophrenia or other psychoses.

There are several possibilities to explain the negativity of these findings. To begin with, the trail may be a false one: the DA-ergic system may have nothing to do with the pathogenesis of psychoses. Secondly, the disorder in the DA system may be selective, i.e., localized in the mesolimbic and mesocortical DA systems. In that case, it would not be measurable: the HVA concentration in CSF reflects chiefly the nigrostriatal DA system; the serum prolactin level, mainly the tubero-infundibular DA system. Another possible explanation can be inferred from the observation of Cools et al. (1975), indicating the existence of two types of DA receptor with opposite functions, in the caudate nucleus as well as in the nucleus accumbens. This would mean that hyperfunction in one of the two systems can arise from two causes: increased availability of DA or increased receptor sensitivity in one system, or hypofunction of the other. The relative hyperfunction would not be associated, one can expect, with increased DA turn-over. A final explanation is that offered by Matthysse (1974). He assumed that DA-ergic hyperactivity is not the primary phenomenon, but secondary to hyper-function of a neuronal network with an inhibitory influence on the DA system. He presented two analogies to support his hypothesis: Parkinson symptoms re-spond favorably to anticholinergic agents. Yet the essential lesion is not cholin-ergic hyperactivity but deterioration of DA neurons. Something similar applies to Huntington's chorea. Even though some symptoms respond favorably to DA antagonists, the DA system is probably not primarily disturbed, but rather the γ-aminobutyric acid system; in Huntington patients, the concentration of this product and of the enzyme which produces it in the basal ganglia is markedly decreased (Bird et al., 1973; Perry et al., 1973).

Matthysse's line of argument fails to convince me. The point is not whether DA-ergic hyperactivity in psychoses is a primary or a secondary phenomenon; it is that there are no findings which convincingly point in the direction of DA-ergic hyperactivity. This situation irrefutably weakens the position of the DA hypothesis.

CHAPTER XX

Testing the DA Hypothesis with the Aid of Drugs

Strategy

If DA-ergic hyperactivity plays a role in the pathogenesis of schizophrenic psychoses, then psychotic symptoms can be expected to be potentiated or induced by substances which increase the amount of DA available at the central DA receptors or enhance the sensitivity of these receptors; inversely, substances which reduce the amount of DA or the sensitivity of the DA receptors can be expected to have antipsychotic properties. In this premise, it is tacitly assumed that an increased availability of transmitter leads to increased neuronal activity in DA-ergic neurons, and vice versa. This is not a certainty, but an accepted starting point in many theories on the relation between brain and behavior.

DA Potentiation by Means of Amphetamines

Amphetamine psychosis. Amphetamines influence the central CA metabolism in a complex manner (Randrup and Munkvad, 1970). They induce release of CA in the synaptic cleft, inhibit their re-uptake into the neuron, and reduce their degradation via MAO inhibition. All these effects lead to an increased transmitter concentration in the CA-ergic synapses and as a result, it is assumed, neuronal activity increases.

Amphetamine derivatives can induce psychoses in nonpsychotic individuals.

This was first described by Young and Scoville (1938) and later studied in detail in 42 patients by Connell (1958). Two types of psychosis are possible: "common" toxic psychosis (usually delirious) and "true" amphetamine psychosis: a paranoid syndrome with ideas of reference, largely auditory hallucinations, ideas of persecution and unclouded, sometimes even unusually lucid consciousness (as it can also be observed in LSD psychosis). The toxic psychosis chiefly develops after one or two very large doses, whereas the "true" syndrome is more likely to develop after repeated parenteral administration. The following focuses exclusively on "true" amphetamine psychosis.

All authors agree that amphetamine psychosis closely resembles paranoid types of schizophrenia. Apart from paranoid symptoms, there is flattening of affect, depressiveness and autistic withdrawal. The difference lies in the fact that, apart from delusions, formal disorders of thinking are rare in amphetamine psychoses. Another difference, of course, lies in the course: amphetamine psychosis is transient and disappears spontaneously after a few days (a course which many authors consider incompatible with a diagnosis of schizophrenia). Kety assumed as early as 1959 that amphetamine psychosis could be a model of certain types of schizophrenia—an assumption later elaborated by Randrup and Munkvad (1972) and Snyder (1973).

Amphetamines in normal volunteers. In normal volunteers with a "clean" history and without schizoid personality features, an amphetamine psychosis can be induced within 1-4 days by giving them increasing doses of amphetamines at short intervals. The amount required is 300-500 mg (Griffith et al., 1968; Angrist et al., 1974). Amphetamine psychosis, therefore, is not simply a latent schizophrenia provoked by drugs. The fact that some test subjects became psychotic after only 24 hours rules out lack of sleep as decisive factor.

Amphetamine psychosis and DA metabolism. There are several indications that DA-ergic hyperstimulation plays a role in the pathogenesis of amphetamine psychosis: (a) Substances which block DA-ergic transmission, such as neuroleptics, are conspicuously quick to produce a favorable effect in amphetamine psychoses (Angrist, 1974), but barbiturates cause no relief (Snyder et al. 1974). (b) Stereotyped behavior, a DA-ergically determined effect of amphetamines in animals (Randrup and Munkvad, 1974), is observed also in amphetamine psychoses (Rylander, 1972). (c) l-Amphetamines and d-amphetamines are about equivalent in their ability to induce psychosis (Angrist et al., 1971); this also applies to their inhibitory effect on (re-)uptake of DA into the neuron (Snyder et al., 1970). But they differ by a factor 10 in their effect on (re-)uptake of NA. This suggests that DA is more important in amphetamine psychosis than NA. (d) In the only test subject so far examined in this respect, amphetamine caused a marked increase in post-probenecid HVA accumulation. This indicates an increased DA turnover. The baseline MHPG concentration (the only indicator of central NA metabolism) remained unchanged (Angrist et al., 1974). (e) Cocaine can induce a psychosis which closely resembles amphetamine psychosis (Mayer-Gross et al., 1960). Since cocaine, too, inhibits (re-)uptake of CA, it is not unlikely that both psychoses are based on the same mechanism.

Amphetamines and schizophrenia. Amphetamines and related stimulants such as methylphenidate (Ritaline) cause aggravation of symptoms in schizophrenic patients even in small doses (Janowsky et al., 1972, 1973). LSD, however, induces in these patients a syndrome which, so to speak, is superimposed on the schizophrenic psychosis. The patient himself recognizes a difference between LSD psychosis and the original syndrome (Hollister, 1962). This effect of amphetamines is not the result of aspecific stimulation of the CNS. Caffeine in large doses has no psychotogenic effect (Angrist, 1974).

In all, amphetamine can be described as a psychosis-inducing compound with a predilection for induction of paranoid symptoms and stimulation of DA receptors probably underlies this effect.

DA Potentiation by Means of l-DOPA

DOPA psychoses. Exogenous l-DOPA is centrally converted largely to DA and but for a small part to NA. As a therapeutic agent in Parkinson's disease, this compound induces psychotic reactions in about 3.5% of cases (Goodwin, 1971; Goodwin et al., 1971). These are of a diverse nature, lack specific characteristics, and show no similarity to "true" amphetamine psychosis. The reasons could be that (a) therapeutic doses of l-DOPA cause a DA increase which is probably much smaller than that after amphetamines in doses which induce psychosis (Snyder et al., 1974), and (b) the DA-producing apparatus is partly destroyed in Parkinson patients.

Furthermore, l-DOPA seems capable of actualizing an existing psychotic predisposition. For example, (hypo)manic reactions after l-DOPA are seen in particular in patients with manic periods in the history and in patients with a history of psychosis the risk of a relapse is increased during DOPA medication (Goodwin, 1972).

DOPA and schizophrenia. l-DOPA has also been given to schizophrenic patients, in daily doses up to 5 g (without peripheral decarboxylase inhibitor). In all cases, this led to clinical deterioration. Most patients showed exacerbation of existing pathology, but some showed only an increased degree of agitation. A few patients developed new symptoms, particularly auditory hallucinations (Angrist et al., 1973, Yaryura-Tobias et al., 1970). Sathananthan et al. (1973) found that the tolerance to the psychosis-inducing effect of l-DOPA is much higher in nonpsychotic psychiatric patients than in schizophrenics. On this ground, they postulated a continuum of threshold values for the amounts of

DA-ergic stimulation which are tolerated without psychotic symptoms—an attractive concept, because it can serve as a kind of biological explanation of the clinical notion that susceptibility to schizophrenic reactions shows marked interindividual variations.

So far as I know, there has been no systematic research into the behavioral effects of large doses of l-DOPA in normal test subjects.

DA Deficiency by Inhibition of Synthesis

α-MT, an inhibitor of tyrosine hydroxylase, the enzyme which catalyzes conversion of tyrosine to DOPA, reduces the amounts of DA and NA available for transmission. If hyperactivity of DA-ergic systems plays a role in the pathogenesis of schizophrenia, then such a substance should have antipsychotic properties. According to Gershon et al. (1967), it has not. In view of the risk of crystalluria and renal damage, however, their dosage was below that at which a significant decrease in CA production can be expected. Carlsson et al. (1972, 1973) therefore studied the question whether α-MT perhaps enhances the effect of neuroleptics. Their procedure was as follows. In a number of schizophrenics stabilized with neuroleptics, they gradually decreased the dosage until the symptoms showed unmistakable aggravation. Next, they combined the neuroleptic with α-MT, up to 2 g daily. If this was insufficient to restore the patient to the initial level, then the dosage of the neuroleptic was increased until this level was attained again. It was found that the neuroleptic dosage could be reduced by an average of 70% (range 30-99%). The extrapyramidal side effects were also potentiated by α-MT—an additional argument to attribute the therapeutic result to CA deficiency. The HVA concentration in the CSF decreased by 70-90%, which indicates partial inhibition of tyrosine hydroxylase activity in the brain.

This experiment does not differentiate between the respective contributions of reduced DA and reduced NA concentration. For this purpose, one should, in principle, resort to an inhibitor of DA-β-hydroxylase, such as fusaric acid (Nagatsu et al., 1970). The combination of a neuroleptic with fusaric acid has not been tested. Fusaric acid as such has been tested, but in manic patients, not in schizophrenics (Sack and Goodwin, 1974). In these patients, its effect was found to be detrimental rather than therapeutic. Psychotic reactions have also been described with the anti-alcoholic agent disulfiram (Antabuse), which, like fusaric acid, is an inhibitor of DA-β-hydroxylase, if a less selective one; this drug is therefore considered contraindicated in schizophrenia (Smythies, 1975). However, it is difficult to interpret these data because, after inhibition of DA-β-hydroxylase, reduced NA production goes hand in hand with increased DA production (all available DOPA now being converted to DA). Any possible antipsychotic effect of the former phenomenon would be abolished by the latter.

Beta-Receptor Block

The clinical (side) effects of neuroleptics are considered to be related to inhibition of transmission in central CA-ergic systems. If this is true, then compounds other than the classical neuroleptics but with a similar biochemical action could be expected to produce a therapeutic effect in psychoses. The only compound which meets these requirements and with which some psychiatric experience has been gained, is propranolol (Inderal): a so-called "β-blocker."

Beta-receptors. On the basis of a pharmacological criterion, the receptors of the peripheral sympathetic nervous system are divided into an α-type and a β-type (for review, see Dollery et al., 1969; Jefferson, 1974). This criterion is: their susceptibility to sympathicomimetic amines. Generally speaking, α-receptors are most susceptible to NA and least susceptible to isoproterenol hydrochloride, whereas the reverse applies to β-receptors, with the exception of intracardiac β-receptors which are also susceptible to NA. The β-receptors are found in abundance in the heart, skeletal muscles, blood vessels and bronchial muscles, and their stimulation leads to an increase in pulse rate, increased myocardial contractility, dilatation of muscular vessels and bronchodilatation. The brain also contains β-receptors which, like those in the heart, are NA-sensitive.

β-blockers inhibit β-receptors (including central β-receptors, at least so far as they enter the brain) in a competitive manner. The degree of β-block at any given moment is therefore a function of the relative concentrations of agonist and antagonist at that moment. Propranolol is a β-blocker capable of entering the CNS, and has no intrinsic sympathicomimetic properties. Some β-blockers combine the ability to block β-receptors with the ability to activate these receptors. Propanolol does not. It does have membrane stabilizing (local anesthetic) properties, but only in doses far in excess of those required to block β-receptors.

Propranolol in psychiatry. Propranolol has been employed in psychiatry for two indications. To begin with, it has been used in nonpsychotic conditions involving anxiety and tension. For this indication, it is certainly effective in that in particular it reduces the somatic symptoms accompanying anxiety and tension (Granville-Grossman, 1974). The peripheral effects of propranolol probably contribute much to this anxiolytic effect.

Secondly, propranolol has been used in psychoses. The first to do so were Atsmon and co-workers, who were put on this trail by a psychotic female suffering from acute porphyria, whom they treated with large doses of propranolol to which she responded by rapid psychological improvement. In two uncontrolled studies (Atsmon et al., 1971, 1972), they treated a small number of schizophrenic patients with large doses of propranolol (average daily dose,

400-4280 mg), and in acute cases obtained striking success within a few days. The more chronic patients showed no or only a much less marked response. All good responders showed an increased renal excretion of CA and MHPG, the latter being the principal metabolite of NA in the CNS, but which can be produced also in the periphery. In a third study, likewise uncontrolled, they compared propranolol with chlorpromazine in 10 women with puerperal psychoses, and found the latter to be equivalent to the former in certain items, and in fact superior in some other items (Steiner et al., 1973). Propranolol in smaller doses and practolol are ineffective in psychoses (Gardos et al., 1973; Rackensperger et al., 1974), but then practolol is a β-blocker which can hardly enter the brain (Scales and Cosgrove, 1970).

Finally, there is a recent study by Yorkston et al. (1975), again inconclusive but certainly indicative of an antipsychotic potency of propranolol. The special feature of this study was that, of the group of 14 schizophrenics studied, 12 were on record as chronic schizophrenics, and none had responded favorably to pheniothiazine-type neuroleptics. Six of the 14 patients showed a complete remission within a week, one showed substantial and two showed moderate improvement. The remaining patients showed no or hardly any improvement. This result would be astonishing, were it not for the many flaws in this study. It was uncontrolled; the concept chronicity was not exactly defined; in some cases, the neuroleptic dosages had been too low to warrant a conclusion of resistance to neuroleptic medication; some patients received propranolol only, while others received propranolol in combination with neuroleptics; and finally, no mention is made of how much extra attention the patients received during the propranolol experiment.

Mechanism of the propranolol effect in psychoses. Provided the propranolol effect can be confirmed in a controlled experiment, what is its mode of action? This question can be divided into three components. First: does the effect originate from the periphery or is it of central origin? Atsmon reported only therapeutic effects in acute patients with an increased CA excretion. Acute psychotic patients often show a marked degree of unrest, anxiety and tension. This might also explain the high CA excretion. In view of this, a peripheral (anxiolytic) effect seems likely. An argument against it, however, is that the Yorkston team observed a therapeutic effect of propranolol also in chronic schizophrenics, in whom unrest and tension are usually less pronounced features.

The second question is whether the antipsychotic effect is based on a local anesthetic or on a β-blocking effect. In the latter case, one would expect a relatively small dose of propranolol (insufficient for a local anesthetic effect) to enhance the therapeutic efficacy of neuroleptics. Research which would supply an answer to this question has not been carried out.

The last question is whether propranolol exclusively blocks NA receptors or also exerts an influence on DA receptors. The former is believed to be true

(Bucher and Schorderet, 1974). By comparing propranolol with a neuroleptic of chiefly DA receptor-blocking potency (e.g., pimozide), one could gain an impression of the relative significance of inhibition of transmission in DA-ergic and NA-ergic systems for the treatment of psychoses.

In all, the propranolol data so far available are insufficient for a proper evaluation, but they amply justify further research. Such studies will have to be made under rigid clinical control in view of the numerous, partly dangerous side effects of large doses of propranolol.

Hallucinogens

Assuming that increased DA-ergic activity plays a role in the pathogenesis of certain psychoses or psychotic symptoms, the question arises whether this mechanism is also involved in so-called "model psychoses"—even though a correlation is not an inevitable conclusion from the hypothesis. It is quite possible, in fact probable, that several different biochemical mechanisms can be involved in the pathogenesis of psychotic manifestations.

We know of only one study which supplies some information on the above question (Bowers, 1972). In 12 patients with an LSD psychosis, the post-pro-benecid HVA accumulation was not abnormal. However, this study was made 1-2 weeks after the last administration of LSD, and therefore warrants no conclusion on a possible acute LSD effect. Moreover, nonpharmacological factors had probably also been active; otherwise, the psychosis would not have persisted so long. The study was not repeated after remission of the psychosis and discontinuation of medication; this means that intra-individual differences between psychotic and nonpsychotic phases cannot be excluded.

Systematic exploration of the influence of hallucinogens of the LSD type on central DA metabolism would be worthwhile as a hypothesis-testing study, but of course meets serious ethical objections for human individuals. Some animal experiments have been carried out in this direction, but the results are confusing. There are indications of an agonistic as well as of an antagonistic activity of LSD in relation to central DA receptors (Pieri et al., 1974; Von Hungen et al., 1974; Smith et al., 1975). There is one indirect indication that DA is of significance for the behavioral effects of LSD. α-MT, an inhibitor of DA and NA synthesis, antagonizes the symptoms of hyperirritability which LSD induces in rabbits (Horita and Hamilton, 1969), and this inhibition is abolished by administration of 1-DOPA combined with a DA-β-hydroxylase inhibitor. The latter prevents conversion of DA to NA (Horita and Hamilton, 1973). Viewed in the context of the DA hypothesis, this is an intriguing finding, but any conclusion would be premature.

Conclusions

The so-called amphetamine psychosis—complication or more or less protracted, excessive and generally parenteral amphetamine use—is a characteristic syndrome which resembles paranoid types of schizophrenia. It can also be induced in normal test subjects. Several observations indicate the likelihood of a hyper-DA-ergic pathogenesis. One fact throws doubt on it: l-DOPA, likewise a DA agonist, can induce psychotic symptoms, but these are variable and generally do not resemble those of classical amphetamine psychosis. However, this is not a conclusive counterargument, because the l-DOPA dosage was such that the increase in DA was probably much less than that attained with amphetamines in psychosis-inducing doses.

Both compounds increase the existing pathology in schizophrenic patients, even if given in relatively modest doses. α-MT, a substance which inhibits CA production and therefore probably the activity in CA-ergic systems, potentiates the therapeutic effect and extrapyramidal side effects of neuroleptics in schizophrenic patients. In all, the pharmacological data seem to support the concept postulated in the DA hypothesis—DA-ergic hyperactivity: a disintegrating factor; inhibition of this hyperactive system: an antipsychotic factor. On the other hand, we have the observation that propranolol has an antipsychotic effect. Propranolol is a β-blocker believed to block central NA receptors, but not DA receptors. This would seem to suggest that the NA-ergic system—as a factor in the pathogenesis of psychoses and in their treatment—has not received the attenton it probably deserves.

CHAPTER XXI

The Predictive Value of the Biochemical Action Profile of Neuroleptics

Chemical Structure vs. Biochemical Mode of Action of Neuroleptics

Groups of compounds of diverse chemical structure behave clinically as neuroleptics. Little is known about differences in clinical effects. Why is agent A prescribed for patient X, and not agent B? The only rational consideration in this respect has so far been the degree of inertia which the patient may show, for some neuroleptics are more likely to aggravate inertia than others. Certain groups are known for their marked sedative effect; others are believed to leave the level of activity and initiative intact, or even slightly to increase it. Yet clinical experience has shown that one cannot generalize about neuroleptics, in this or in any other respects. In a given patient, one neuroleptic may be ineffective, and the other efficacious. Some patients are resistant to neuroleptics, while other patients with a similar psychopathology respond favorably. In this way, neuroleptic medication entails many uncertainties, which their chemical structure cannot explain.

This situation poses two questions: (a) Could the biochemical action profile of neuroleptics, i.e., the degree to which they block DA-ergic or NA-ergic transmission, perhaps be a better predictor of clinical efficacy than their chemical structure? This would be in the line of expectation, assuming that these biochemical actions are indeed related to the clinical activity. (b) Can biochemical variables be found in psychotic patients, and in particular with regard to the central CA metabolism, with a predictive value as to the clinical effect?

A limited number of studies have been devoted to the first question; the second question is still entirely virgin country.

Biochemical vs. Clinical Action Profile of Neuroleptics

Chlorpromazine versus oxypertine. Van Praag et al. (1975b) made a comparative study of the clinical efficacy of chlorpromazine (Thorazine) and oxypertine (Opertil) in patients with acute psychoses of diverse etiology and symptomatology. The choice of these two compounds was determined by biochemical considerations.

Chlorpromazine increases the turnover of DA and NA in the brain in test animals: an indication that DA as well as NA receptors are blocked (Andén et al., 1970). The human central DA turnover also increases, but the NA metabolism is not demonstrably influenced, this, however, may be due to the method used (Van Praag and Korf, 1975). Because probenecid does not inhibit MHPG transport from the CNS (Korf et al., 1971), we must rely for information on the central NA metabolism on determination of the baseline concentration of MHPG in the CSF, and baseline concentrations of MA metabolites are a less faithful reflection of the metabolism of the mother amines than the postprobenecid concentrations (chapter X).

Oxypertine is a neuroleptic of a different type (Hassler et al., 1970). It does not block CA receptors but causes depletion of intraneuronal CA stores, and in this respect it is fairly selective. Within a given dosage range, oxypertine has a predilection for NA stores: the NA concentration decreases, that of the NA metabolites increases, but the DA metabolism is but little influenced. Its activity in human individuals is probably similar (Van Praag and Korf, 1975). A substance which more or less selectively blocks central NA receptors was not available at the time, and still is not. The two substances were selected because chlorpromazine reduces both DA-ergic and NA-ergic transmission, while oxypertine reduces mainly NA-ergic transmission. On the basis of the biochemical action of these neuroleptics, the following hypotheses were tested.

(1) Chlorpromazine is superior to oxypertine as an overall therapeutic agent in psychoses. The difference is based on a more pronounced therapeutic effect of chlorpromazine on delusions and hallucinations.

Motivation: an increase in central DA-ergic activity may be a psychosis-provoking factor. The principal indication that this is so is that amphetamines and l-DOPA can induce psychotic symptoms in normal individuals and cause exacerbation of these symptoms in psychotic patients, while both drugs increase

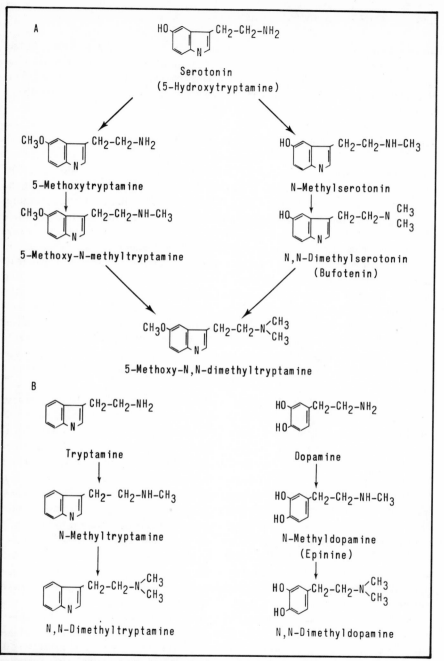

Methylation reactions of naturally occurring biogenic amines mediated by 5-methyl THF: (A) Serotonin can be both O-methylated and N-methylated; (B) N-methylation of dopamine and tryptamine (from Snyder, 1974).

illusory. The chance of identifying them in CSF would seem to be somewhat better, although even this will probably be impossible without gas chromatography and mass spectrometry—the most sensitive methods to separate and identify such metabolites. In this manner, however, the dysmethylation hypothesis has not yet been tested.

Pharmacological Testing of the Dysmethylation Hypothesis

Methionine and dysmethylation. To verify the dysmethylation hypothesis, Pollin et al. (1961) administered large doses of l-methionine to chronic schizophrenics, who had all been premedicated with a MAO inhibitor. In the organism, including the brain, methionine is converted to S-adenosylmethionine—a methyl donor (Baldessarini, 1966). The basic hypothesis was the expectation that an increase in unstable methyl groups would facilitate trans(dys)methylation and cause exacerbation of schizophrenic symptoms. The MAO inhibitor was given in order to delay the degradation of the abnormal methylation products. Indeed, 40% of the patients thus treated showed deterioration. This was believed not to involve a superimposed toxic psychosis but really to represent exacerbation of the schizophrenic symptoms. Methionine alone had no effect on behavior. This observation has been corroborated in several double-blind studies (Park et al., 1965; Berlet et al., 1965; Narasimhachari et al., 1970), with the restriction that disorders of consciousness did occur according to some investigators—a fact reminiscent after all of a toxic component.

The possible role of the MAO inhibitor has been somewhat ignored in these experiments. These compounds can produce psychotic reactions (Crane, 1956), and only one study showed that methionine alone, if given in large doses (10-20 g/day) can cause aggravation of symptoms (Antun et al., 1971). On the other hand, a few nonschizophrenic psychiatric patients and normal test subjects showed no or only a minimal reaction to the methionine/MAO inhibitor combination. This demonstrates that schizophrenic patients are at least unusually sensitive to it, although of course this does not prove that the pathogenetic process of schizophrenia is really directly activated. Another argument in favor of a relation between (dys)methylation and schizophrenia was the observation reported by Hall et al. (1969) that COMT, the enzyme which transports the methyl group of a methyl donor to the CA, produced exacerbation in schizophrenic patients. These were uncontrolled studies, however, and the observations have not been corroborated.

How could hypermethylation of MA occur, always assuming that it is indeed a schizophrenia-inducing mechanism? Laduron (1974b) considered as a possibility a defect in the degradation of the methyl donor 5-methyl-THF, resulting in an excess of unstable methyl groups and increased (dys)methylation (chapter XIX). In support of this possibility, he quoted unpublished observations which were said to indicate that folic acid, precursor of 5-methyl-THF, aggravates schizophrenic symptoms. These data still await confirmation.

Other explanations of the methionine effect. The methionine effect in schizophrenia can also be explained in other ways. S-adenosylmethionine, a product to which methionine is converted, inhibits methylene-THF reductase, which is the enzyme involved in the synthesis of 5-methyl-THF (Kutsbach and Stokstad, 1967). In intracerebral methylation reactions, not S-adenosylmethionine but 5-methyl-THF was assumed to be the principal methyl donor. According to this theory, methionine would reduce rather than increase the supply of unstable methyl groups suitable for methylation of MA. As a result, MA degradation stagnates, including DA degradation, and DA accumulation results, which brings us back to the DA hypothesis.

Another possibility is that the methionine effect is related to a deficiency in another amino acid, e.g., l-tryptophan, due to interference of the excess l-methionine with their uptake into the neuron.

Tryptophan. Of the other amino acids studied, only tryptophan (combined with a MAO inhibitor) caused exacerbation of schizophrenia in some cases (Pollin et al., 1961). In normal test subjects without a history of psychoses and premedicated with a MAO inhibitor, a single large dose of l-tryptophan (5 g orally) can induce a psychosis of an LSD-like character: delusion-like ideas occur, time perception and perception of the external world become distorted, and consciousness remains unclouded (Van Praag, 1962). The cause of the psychosis is a matter of speculation. Large amounts of tryptophan stimulate central 5-HT and tryptamine production, and an excess of these substances cannot be readily handled as a result of the MAO inhibition. Under these circumstances, it is conceivable that abnormal methylation of 5-HT or tryptamine occurs or that a physiological methylation process assumes abnormal proportions. CSF studies could clinch this theory. Do methyl derivatives of IA appear, and is their appearance related to a possible development of psychiatric symptoms? No such research has been carried out.

Treatment with nicotinic acid. Spectacular success has been reported in schizophrenia with nicotinic acid, an acceptor of methyl groups (Hoffer, 1962). This would be a strong argument in favor of the dysmethylation hypothesis, were it not that these results are as yet to be confirmed (Ban and Lehmann, 1970). On the other hand, it has been found that nicotinic acid has no effect whatever on the concentration of S-adenosylmethionine in the rat brain (Baldessarini, 1966), and this removes the rationale for prescribing this compound in schizophrenia.

Conclusions

The dysmethylation hypothesis is an attractive possibility, but no more than that. Some methyl derivatives of IA and DA have hallucinogenic properties. The human brain contains enzymes capable of effecting such methylations. However, the question whether these occur *in vivo* or play a role in the pathogenesis of psychoses is still moot.

There are reports on hallucinogenic 5-HT and DA derivatives in urine and blood from schizophrenic patients, but they are controversial, mainly because the chemical methods used do not ensure unequivocal separation and identification of the substances in question. Verification of the dysmethylation hypothesis will require gas chromatography and mass spectrometry, and in the present situation I would have more confidence in CSF studies than in findings obtained in peripheral body fluids.

Snyder et al. (1974) rightly pointed out that the methylation process of MA must not in advance be exclusively related to psychopathology. Hallucinogenic methyl derivatives of 5-HT and DA, like LSD and large doses of amphetamine derivatives, can induce a psychosis in which unclouded consciousness and increased susceptibility to endogenous and exogenous stimuli are important features. These symptoms sometimes occur without any sign of psychosis. The user then speaks of an expansion of consciousness, not without justification. The perception threshold can, so to speak, be adjusted at different levels. It is an attractive working hypothesis that methyl derivatives of IA and DA are involved in this process.

General Conclusions

(1) Research into the biological determinants of schizophrenic psychoses has long suffered from a lack of well-founded and testable working hypotheses. The neuroleptics have changed this situation. This course of events can be compared with that in biological depression research, to which the antidepressants have given strong impetus. Neuroleptics are substances which, although heterogeneous in terms of chemical structure, show two similarities. Biochemically: they reduce DA-ergic (and NA-ergic) transmission in the brain. Psychopathologically: they reduce unrest and psychotic disorders of thinking and experiencing. These facts prompt two questions, which together provide a research strategy. The first is whether neuroleptics influence the human central CA metabolism as they do in animals. The second is whether DA-ergic (and perhaps also NA-ergic) hyperactivity plays a role in the pathogenesis of schizophrenic psychoses. The so-called "DA hypothesis" postulates such a relation.

Another research line derives from the knowledge that O-methylation and N-methylation of biogenic amines yields compounds with a hallucinogenic potency. This raises the question whether these methylations can occur *in vivo*, and, if so, whether they play a role in the pathogenesis of psychotic symptoms. This is postulated in the so-called "transmethylation hypothesis".

(2) The hypothesis that the clinical (side) effects of neuroleptics of the DA receptor-blocking type correlate with their influence on the central DA metabolism is a fairly sound one. In human individuals and test animals alike, these compounds increase the central DA turnover (taking as a yardstick the probenecid-induced increase in the HVA concentration in lumbar CSF), and this effect is probably secondary to *inhibition* of transmission, *not* an expression of *increased* neuronal activity. Moreover, there is a positive correlation between the degree of increase in DA turnover in response to neuroleptics, and the intensity of their therapeutic effect. The hypokinetic-rigid symptoms, too, are as much more pronounced as the DA turnover is more markedly increased, and the risk of occurrence of these side effects is greater at a high than at a low pre-therapeutic DA turnover.

The neuroleptics in question block not only DA-ergic but also NA-ergic transmission. The significance of this fact for their clinical effects is rather obscure. There are indications that propranolol, a β-blocker which blocks central postsynaptic NA receptors but leaves DA receptors uninfluenced, can have a therapeutic effect in schizophrenic psychoses. In view of this, there are sound reasons to advocate further investigation of the NA-ergic system in its relation to the clinical effect of neuroleptics.

(3) The biochemical action profile of neuroleptics seems to be a more reliable predictor of their clinical (side) effects than their chemical structure. This was to be expected if the influence of these compounds on the central CA metabolism does indeed underlie (some of) their therapeutic effects. This implies that efforts to develop new neuroleptics should focus on the biochemical action of the compounds tested, more than has so far been the case. In this context, I define the designation "biochemical action" as the extent to which these agents suppress DA-ergic and NA-ergic transmission, and their relative influence on the nigrostriatal, mesolimbic and mesocortical DA systems. It appears to me that this approach opens wider perspectives than any random varying of the chemical structure of known neuroleptic substances.

(4) Only indirect arguments can be marshaled in favor of the hypothesis that hyperactivity of central DA-ergic systems plays an important role in the pathogenesis of schizophrenic psychoses. Drugs which can be considered to increase the activity of these systems, specifically amphetamines and l-DOPA, provoke psychoses and aggravate schizophrenic symptoms. The therapeutic effect of neuroleptics is potentiated by substances which inhibit DA (and NA) synthesis. Direct arguments, however, can hardly be found: it is true that the central DA turnover can be increased in schizophrenic (and other) psychoses, but this phenomenon is more likely to be related to increased motor activity than to prototypical psychotic symptoms such as delusion and hallucination. The normal serum prolactin levels, too, are not suggestive of increased central DA activity.

There are some indications that CA degradation and conversion of DA to NA are disturbed in schizophrenic patients, but no unequivocal evidence in support of these indications has so far been advanced.

This is no reason to reject the DA hypothesis. The HVA concentration in CSF reflects mainly the DA metabolism in the nigrostriatal DA system, whereas serum prolactin is a function of the activity in the tubero-infundibular DA system. There are no methods which give information on the human mesolimbic and mesocortical DA systems.

(5) Moreover, the contention that psychiatric syndromes or diseases might be pathogenetically determined by one well-defined biochemical lesion is probably an illusion. It seems much more likely to me that several different biochemical functional disorders underlie psychiatric syndromes, and that each of these disorders is responsible for certain features of the syndrome. Whenever a biochemical phenomenon is not nosologically or at least syndromally specific, it is often dismissed as insignificant—without justification. The concept of a symptomatological specificity of biological variables is tenable (in this context, I define a symptom as a well-defined disturbance in a psychological function or series of interrelated psychological functions).

(6) O-methylation and N-methylation of 5-HT, DA and tryptamine—biogenic amines which occur in the brain—produces compounds with an LSD-like hallucinogenic activity. The human brain contains enzymes which, in principle, are capable of such methylations. Whether indeed they do so under certain circumstances, and whether this comprises a mechanism in the pathogenesis of (schizophrenic) psychoses, remains uncertain. There are no convincing indications that the CSF contains metabolites of this type during psychotic phases, but then the most adequate chemical technology—gas chromatography combined with mass spectrometry—has not so far been applied to this problem. The question whether methylated MA metabolites are involved in the regulation of normal psychological functions, e.g., the degree of susceptibility to exogenous stimuli, has not been studied at all.

(7) The group of schizophrenic psychoses is a heterogeneous collection of syndromes, which have in common that they usually involve no clouding of consciousness, but which cannot otherwise be brought under a common denominator, either in symptomatological, etiological or prognostic terms. It is therefore an absolute necessity to ensure consistent classification of these syndromes according to three criteria: symptomatology, etiology and course (a procedure which indeed has great advantages for psychiatric classification in general).

Even after such a three-dimensional diagnosis, however, it remains difficult to compose test groups which tend toward homogeneity within a diagnostic category. This is why longitudinal studies, with each patient serving as his own control, are advisable. If a given biochemical disorder occurs in 10% of a group of patients diagnosed as schizophrenic, and if this phenomenon covaries with the psychological condition, then it provides an indication that one is dealing with a real subgroup. If the biochemical results in this probable subgroup had simply been considered in lump with those in the remaining 90% of patients, then the differentiation would have been effaced.

(8) Accurate psychopathological classification of schizophrenic psychoses is indispensable in the study of their biological substrates. Inversely, biological research can supply indications useful in differentiating subgroups which differ in such factors as pathogenetic mechanism, response to therapy, and prognosis. This means that biological schizophrenia research can be meaningful only if it is carried out in correlation with psychopathological research.

References

Ackenheil, M., Beckman, H., Hoffmann, G., Markianos, E., Nyström, I., and Raese, J. (1974): Einflusz von Clozapin auf die MHPG-, HVS- und 5-HIES-Ausscheidung im Urin und liquor cerebrospinalis. *Arzneimittelforschung, 24:* 984-987.

Alousi, A., and Weiner N. (1966): The regulation of norepinephrine synthesis in sympathetic nerves: effect of nerve stimulation, cocaine and catecholamine-releasing agents. *Proceedings of the National Academy of Sciences, 56:* 1491-1496.

Andén, N.E., Roos, B-E., and Werdinius B. (1964): Effects of chlorpromazine, haloperidol and reserpine on the levels of phenolic acids in rabbit corpus striatum. *Life Sciences, 3:* 149-158.

Andén, N.E., Rubenson, A., Fuxe, K., and Hökfelt, T. (1967): Evidence for dopamine receptor stimulation by apomorphine. *Journal of Pharmacy and Pharmacology, 19:* 627-629.

Andén, N.E., Butcher, S.G., Corrodi, H., Fuxe, K. and Ungerstedt, U. (1970a): Receptor activity and turnover of dopamine and noradrenaline after neuroleptics. *Journal of Pharmacology, 11:* 303-314.

Andén, N.E., Corrodi, H., Fuxe, K., Hökfelt, T., Hökfelt, C., Rydin, C., and Svensson, T. (1970b): Evidence for a central noradrenaline receptor stimulation by clonidine. *Life Sciences, 9:* 513-523.

Andén, N.E., and Bédard, P. (1971): Influences of cholinergic mechanisms on the function and turnover of brain dopamine. *Journal of Pharmacy and Pharmacology, 23:* 460-462.

Andén, N.E., Corrodi, H., and Fuxe, K. (1972): Effect of neuroleptic drugs on central catecholamine turnover assessed using tyrosine- and dopamine-β-hydroxylase inhibitors. *Journal of Pharmacology, 24:* 177-182.

Andén, N.E., and Stock, G. (1973): Effect of clozapine on the turnover of dopamine in the corpus striatum and in the limbic system. *Journal of Pharmacy and Pharmacology, 25:* 346-348.

Angrist, B.M., Shopsin, B., and Gershon, S. (1971): The comparative psychotomimetic effects of stereo-isomers of amphetamine. *Nature, 234:* 152-153.

Angrist, B.M., Sathanathan, G., and Gershon, S. (1973): Behavioral effects of L-DOPA in schizophrenic patients. *Psychopharmacologia, 31:* 1-12.

Angrist, B.M., Sathanathan, G., Wilk, S., and Gershon, S. (1974a): Behavioral and biochemical effects of L-DOPA in psychiatric patients. In: Frontiers in catecholamine research. Edited by E. Usdin and S.H. Snyder. Pergamon Press, New York. Pp. 991-994.

Angrist, B.M., Sathanathan, G., Wilk, S., and Gershon, S. (1974b): Amphetamine psychosis: behavioral and biochemical aspects. *Journal of Psychiatric Research, 11:* 13-23.

Antun, F.T., Burnett, G.B., Cooper, A.J., Daly, R.J., Smythies, J.R., and Zeally, A.K. (1971): The effects of L-methionine (without MAOI) in schizophrenia. *Journal of Psychiatric Research, 8:* 63-71.

Arieti, S. (1974): *Interpretation of Schizophrenia.* Basic Books, New York.

Atsmon, A., Blum, I., Maoz, B., Steiner, M., Ziegelman, G., and Wijsenbeek, H. (1971): The short-term effects of adrenergic blocking agents in a small group of psychotic patients: preliminary clinical observations. *Psychiatria, Neurologia, Neurochirurgia, 74:* 251-258.

Atsmon, A., Blum, I., Steiner, M., Latz, A., and Wijsenbeek, H. (1972): Further studies with propranolol in psychotic patients. *Psychopharmacologia, 27:* 249-254.

Axelrod, J. (1957): O-methylation of epinephrine and other catechols in vitro and in vivo. *Science, 126:* 400-401.

Axelrod, J. (1961): Enzymatic formation of psychotomimetic metabolites from normally occuring compounds. *Science, 134:* 343-344.

Ayhan, I.H., and Randrup, A. (1973): Behavioural and pharmacological studies on morphine-induced exitation of rats. Possible relation to brain catecholamines. *Psychopharmacologia* (Berl.) *29:* 371-329.

Baldessarini, R.J. (1966): Factors influencing tissue levels of the major methyl donor in mammalian tissue. In: *Amine Metabolism in Schizophrenia.* Edited by H.E. Himwich, S.S. Kety and J.R. Smythies. Pergamon Press, New York. Pp. 199-207.

Ban, T.A., and Lehmann, H.E. (1970): Nicotinic acid in the treatment of schizophrenics. In: Canadian Mental Health Association Study. Progress Report I. Canadian Mental Health Association, Toronto.

Banerjee, S.P., and Snyder, S.H. (1973): Methyltetrahydrofolic acid mediates N- and O-methylation of biogenic amines. *Sciences, 182:* 74-75.

Barbeau, A. (1972a): Dopamine and mental function. In: *L-DOPA and Behavior.* Edited by S. Malitz. Raven Press, New York.

Barbeau, A. (1972b): Role of dopamine in the nervous system. Monographs in human Genetics, 6: 114-130.

Barolin, G.S., Bernheimer, H., and Hornykiewicz, O. (1964): Seitenverschiedenes Verhalten des Dopamines (3-Hydroxytyramin) im Gehirn eines Falles von Hemiparkinsonismus. *Schweizer Archiv für Neurologie und Psychiatrie, 94:* 241-248.

Bartholini, G., and Pletscher, A. (1971): Atropine-induced changes of cerebral dopamine turnover. *Experientia, 27:* 1302-1303.

Bartholini, G., and Pletscher, A. (1972): Drugs affecting monoamines in the basal ganglia. In: *Studies of Neurotransmitters at the Synaptic Level.* Edited by E. Costa, L.L. Iversen and R. Paoletti. Raven Press, New York. Pp. 135-148.

Bein, H.J. (1956): The pharmacology of rauwolfia. Pharmacological Reviews, 8: 435-483.

Berlet, H.H., Mutsumoto, K., Pscheidt, G.R., Spaide, J., Bull, C. and Himwich, H.E. (1965):

Biochemical correlates of bheavior in schizophrenic patients. Archives of General Psychiatry, 13: 521-531.

Bird, E.D., MacKay, A.V.P., Rayner, C.N., and Iversen, L.L. (1973): Reduced glutamic-acid decarboxylase activity of post-mortem brain in Huntington's Chorea. *Lancet, I,* 1090-1092.

Bleuler, E. (1911): Dementia Praecox oder die Gruppe der Schizophrenien. In: *Handbuch der Psychiatrie.* Edited by G. Aschaffenburg. Euticke, Leipzig.

Bleuler, E. (1923): *Lehrbuch der Psychiatrie.* Springer, Berlin.

Bowers, M.B., Jr., Heninger, G.R., and Gerbode, F. (1969): Cerebrospinal fluid 5-hydroxyindoleacetic acid and homovanillic acid in psychiatric patients. *International Journal of Neuropharmacology, 8:* 255-262.

Bowers, M.B., Jr. (1972): Acute psychosis induced by psychotomimetic drug abuse. II. Neurochemical findings. *Archives of General Psychiatry, 27:* 440-442.

Bowers, M.B., Jr. (1973): 5-Hydroxyindoleacetic acid (5-HIAA) and homovanillic acid (HVA) following probenecid in acute psychotic patients treated with phenothiazines. *Psychopharmacologia, 28:* 309-318.

Bowers, M.B., Jr. (1974): Central dopamine turnover in schizophrenic syndromes. *Archives of General Psychiatry, 31:* 50-54.

Bowers, M.B., Jr., and Rozitis, A. (1974): Regional differences in homovanillic acid concentrations after acute and chronic administration of antipsychotic drugs. *Journal of Pharmacy and Pharmacology, 26:* 743-745.

Bowers, M.B., Jr. (1976): Fluorometric measurement of 5-hydroxyindoleacetic acid (5-HIAA) and tryptophan. *Biological Psychiatry* (in press).

Bruno, A., and Cumer Bruno, S. (1966): Effects of L-DOPA on Pharmacological parkinsonism. *Acta Psychiatrica Scandinavica, 42:* 264-271.

Bucher, M.B., and Schorderet, M. (1974): Apomorphine-induced accumulation of cyclic AMP in isolated retinas of the rabbit. *Biochemical Pharmacology, 23,* 3079-3082.

Bunney, B.S., Jr., Walters, J.R., Roth, R.H., and Aghajanian, G.K. (1973): Dopaminergic neurons: effect of antipsychotic drugs and amphetamine on single unit activity. *Journal of Pharmacology and Experimental Therapeutics, 185:* 560-571.

Carenzi, A., Gillin, J.C., Guidotti, A., Schwartz, M.A., Trabucchi, M., and Wyatt, R.J. (1975): Dopamine-sensitive adenylyl cyclase in human caudate nucleus. A study in control subjects and schizophrenic patients. *Archives of General Psychiatry, 32:* 1056-1059.

Carlsson, A., and Lindqvist, M. (1963): Effect of chlorpromazine or haloperidol on formation of 3-methoxytyramine and normetanephrine in mouse brain. *Acta Pharmacologica (Kbh), 20:* 140-144.

Carlsson, A., Persson, T., Roos, B-E. and Wålinder, J. (1972): Potentiation of phenothiazines by α-methyl-tyrosine in treatment of chronic schizophrenia. *Journal of Neural Transmission, 33:* 83-90.

Carlsson, A., Roos, B-E., Wålinder, J. and Skott, A. (1973): Further studies on the mechanism of antipsychotic action: potentiation by α-methyl-tyrosine of thioridazine effects in chronic schizophrenics. *Journal of Neural Transmission, 34:* 125-132.

Carlsson, A. (1974): Antipsychotic drugs and catecholamine synapses. *Journal of Psychiatric Research, 11:* 57-64.

Carpenter, W.T., Strauss, J.S., and Muleh, S. (1973): Are there pathognomonic symptoms in schizophrenia? *Archives of General Psychiatry, 28:* 847-852.

Carpenter, W.T., Murphy, D.L., and Wyatt, R.J. (1975): Platelet monoamine oxidase activity in acute schizophrenia. *American Journal of Psychiatry, 132:* 438-441.

Chase, T.N., Schnur, J.A., and Gordon, E.K. (1970): Cerebral spinal fluid monoamine catabolites in drug-induced extrapyramidal disorders. *Neuropharmacology, 99:* 265-268.

Chase, T.N. (1972): Drug-induced extrapyramidal disorders. In: *Neurotransmitters*. Edited by I.J. Kopin. Williams and Wilkins, Baltimore. Pp. 448-471.

Clement-Cormier, Y.C., Kebabian, J.W., Petzhold, G.L., and Greengard, P. (1974): Dopamine sensitive adenylate cyclase in mammalian brain: a possible site of action of antipsychotic drugs. *Proceedings of the National Academy of Sciences, U.S.A., 71:* 1113-1117.

Collaborative Study Group, U.S.A. (Cole, J.O., et al.) (1964): Phenothiazine treatment in acute schizophrenia. *Archives of General Psychiatry, 19:* 246-261.

Connell, P.H. (1958): *Amphetamine Psychosis*. Maudsley Monographs, no. 5. Oxford University Press, London.

Cools, A.R. (1973): The caudate nucleus and neurochemical control of behaviour. The function of dopamine and serotonine in the caput nuclei caudati of cats. Thesis, Nijmegen.

Cools, A.R. (1975): An integrated theory of the etiology of schizophrenia: impairment of the balance between certain, in series connected dopaminergic, serotonergic, and noradrenergic pathways within the brain. In: *On the Origin of Schizophrenic Psychoses*. Edited by H.M. van Praag. De Erven Bohn B.V., Amsterdam. Pp. 58-80.

Costa, E., and Neff, N.H. (1966): Isotopic and non-isotopic measurements of the rate of catecholamine biosynthesis. In: *Biochemistry and Pharmacology of the Basal Ganglia*. Edited by E. Costa, L.J. Coté and M.D. Yahr. Raven Press, New York. Pp. 141-156.

Costall, B., and Naylor, R.J. (1975): The role of the raphé and extrapyramidal nuclei in the stereotyped and circling responses to quipazine. *Journal of Pharmacy and Pharmacology, 27:* 368-371.

Cotzias, G.C., Van Woert, M.H., and Schiffer, L.M. (1967): Aromatic amino acids and modification of parkinsonism. *New England Journal of Medicine, 276:* 374-379.

Crane, G.E. (1956): Further studies on iproniazid phosphate. *Journal of Nervous and Mental Disease, 124:* 322-331.

Crow, T.J., and Gillbe, C. (1973): Dopamine antagonism and antischizophrenic potency of neuroleptic drugs. *Nature/New Biology, 245:* 27-28.

Da Prada, M., Saner, A., Burkard, W.P., Bartholini, G., and Pletscher, A. (1975): Lysergic acid diethylamine: evidence for stimulation of cerebral dopamine receptors. *Brain Research, 94:* 67-73.

Davis, J.M. (1974): Dose equivalence of the antipsychotic drugs. *Journal of Psychiatric Research, 11:* 65-69.

Dijk, W.K. van (1963): *Psychopathologische en klinische aspecten van de psychogene psychose*. Dissertatie, Groningen.

Dollery, C.T., Paterson, J.W., and Conolly, M.E. (1969): Clinical pharmacology of beta-receptor-blocking drugs. *Clinical Pharmacology and Therapeutics, 10:* 765-799.

Donoso, A.O., Bishop, W., Fawcett, C.P., Krulich, L., and McCann, S.M. (1971): Effect of drugs that modify brain monoamine concentrations on plasma gonadotropin and prolactin levels in the rat. *Endocrinology, 89:* 774-784.

Dunner, D.L., Cohn, C.K., Weinshiboum, R.M., and Wyatt, R.J. (1973): The activity of dopamine-β-hydroxylase and methionine-activating enzyme in blood of schizophrenic patients. *Biological Psychiatry, 6:* 215-220.

Editorial (1975): Family deviance and schizophrenia. *Lancet, II:* 213-214.

Faergeman, P.M. (1963): *Psychogenic Psychoses*. Butterworth & Co., London.

Falek, A., and Moser, H.M. (1975): Classification in schizophrenia. *Archives of General Psychiatry, 32:* 59-67.

Freeman, J.M., Finkelstein, J.D., and Mudd, S.H. (1975): Folate-responsive homocystinuria and "schizophrenia." *New England Journal of Medicine, 292:* 491-496.

Friedhoff, A.J., and Van Winkle, E. (1962): Isolation and characterization of a compound from the urine of schizophrenics. *Nature, 194:* 867-869.

Fujimori, M., and Alpers, H.S. (1971): Psychotomimetic compounds in man and animals. In: *Biochemistry, Schizophrenia and Affective Disorders*. Edited by H.E. Himwich. Williams and Wilkins, Baltimore. Pp. 361-413.

Fuxe, K., and Ungerstedt U. (1970): Histochemical, biochemical and functional studies on central monoamine neurons after acute and chronic amphetamine administration. In: *Amphetamines and Related Compounds*. Edited by E. Costa and S. Garattini. Raven Press, New York. Pp. 257-288.

Gardos, G., Cole, J.O., Orzack, M.H., and Volicer, L. (1973): Propranolol in treatment-resistant schizophrenics. *Psychopharmacology Bulletin, 9:* 43-44.

Garelis, E., Gillin, J.C., Wyatt, R.J., and Neff, N. (1975): Elevated blood serotonin concentrations in unmedicated chronic schizophrenic patients: a preliminary study. *American Journal of Psychiatry, 132:* 184-186.

Gerbode, F.A., and Bowers, M.B., Jr. (1968): Measurement of acid monoamine metabolites in human and animal cerebrospinal fluid. *Journal of Neurochemistry, 15:* 1053-1055.

Gerlach, J., Thorsen, K., and Fog, R. (1975): Extrapyramidal reactions and amine metabolites in cerebrospinal fluid during haloperidol and clozapine treatment of schizophrenic patients. *Psychopharmacologia, 40:* 341-350.

Gershon, S., Hekimian, L.J., Floyd, A., Jr. and Hollister, L.E. (1967): Methyl-p-tyrosine (AMT) in schizophrenia. *Psychopharmacologia, 11:* 189-194.

Gershon, S., and Shader, R.I. (1969): Screening for aminoacidurias in psychiatric inpatients. *Archives of General Psychiatry, 21:* 82-88.

Gessner, P.K., and Page, I.H. (1962): Behavioral effects of 5-methoxy-N,N-dimethyltryptamine, other tryptamines and LSD. *American Journal of Physiology, 203:* 167-172.

Goodwin, F.K. (1971): Psychiatric side effects of L-DOPA in man. *JAMA, 218:* 1915-1920.

Goodwin, F.K., Murphy, D.L., Brodie, H.K., and Bunney, W.E. (1971): Levodopa: alterations in behavior. *Clinical Pharmacology and Therapeutics, 12:* 383-396.

Goodwin, F.K. (1972): Behavioral effects of L-DOPA in man. In: *Psychiatric Complications of Medical drugs*. Edited by R.I. Shader. Raven Press, New York, pp. 149-174.

Gordon, M. (1967): Phenothiazines. In: *Psychopharmacological Agents*. Edited by M. Gordon. Academic Press, New York. Pp. 2-198.

Granville-Grossman, K. (1974): Propranolol anxiety and the central nervous system. *British Journal of Clinical Pharmacology, 1:* 361-363.

Green, A.R., Koslow, S.H., and Costa, E. (1973): Identification and quantitation of a new indolealkylamine in rat hypothalamus. *Brain Research, 51:* 371-374.

Griffith, J.J., Oates, J., and Cavanaugh, J. (1968): Paranoid episodes induced by drugs. *Journal of the American Medical Association, 205:* 39-44.

Hall, P., Hartridge, G., and Leeuwen, G.H. van (1969): Effect of catechol O-methyl transferase in schizophrenia. *Archives of General Psychiatry, 20:* 573-575.

Hassler, R., Bak, I.J., and Kim, J.S. (1970): Unterschiedliche Entleerung der Speicherorte für Noradrenalin, Dopamin und Serotonin als Wirkungsprinzip des Oxypertins. *Nervenarzt, 41:* 105-118.

Hecker, E. (1871): Die Hebephrenie. *Archiv für Pathologische Anatomie und Physiologie und für Clinishe Medizin, 52:* 394-429.

Hoffer, A., Osmond, H., and Smythies, J. (1954): Schizophrenia: a new approach. *Journal of Mental Sciences, 100:* 29-54.

Hoffer, A. (1962): *Niacin Therapy in Psychiatry*. Charles C. Thomas, Springfield.

Hollister, L.E. (1962): Drug-induced psychoses and schizophrenic reactions: critical comparison. *Annals of the New York Academy of Sciences, 96:* 80-92.

Horita, A., and Hamilton, A.E. (1969): Lysergic acid diethylamide: Dissociation of its behavioral and hyperthermic actions by D,L-α-methyl-p-tyrosine. *Science, 164:* 78-79.

Horita, A., and Hamilton, A.E. (1973): The effects of D,L-α-methyl-p-tyrosine and L-DOPA on the hyperthermic and behavioral actions of LSD in rabbit. *Neuropharmacology, 12:* 471-476.

Horn, A.S., Cuello, A.C., and Miller, R.J. (1974): Dopamine in the mesolimbic system of the rat brain: endogenous levels and the effect of drugs on the uptake mechanism and stimulation of adenylate cyclase activity. *Journal of Neurochemistry, 22:* 265-270.

Hornykiewicz, O. (1966): Dopamine (3-hydroxytyramine) and brain function. *Pharmacological Reviews, 18:* 925-964.

Hornykiewicz, O. (1972): Dopamine and extrapyramidal motor function and dysfunction. In: *Neurotransmitters.* Edited by I.J. Kopin. Williams and Wilkins, Baltimore. Pp. 390-415.

Hungen, K. van, Roberts, S., and Hill, D.F. (1974): LSD as an agonist and antagonist at central dopamine receptors. *Nature, 252:* 588-589.

Iversen, L.L. (1975): Dopamine receptors in the brain. *Science, 188:* 1084-1089.

Janowski, D.S., El-Yousef, M.K., and Davis, J.M. (1972): The elicitation of psychotic symptomatology by methylphenidate. *Comprehensive Psychiatry, 13:* 83.

Janowski, D.S., and Sekerke, H.J. (1973): Parasympathetic suppression of manic symptoms by physostigmine. *Archives of General Psychiatry, 28:* 542-552.

Jefferson, J.W. (1974): Beta-adrenergic receptor blocking drugs in Psychiatry. *Archives of General Psychiatry, 31:* 681-691.

Kahlbaum, K.L. (1874): *Clinische Abhandlungen einige Psychische Krankheiten. I. Katatonia oder das Spannungsirresein.* Springer, Berlin.

Kamberi, I.A., Mical, R.S., and Porter, J.C. (1971a): Effect of anterior pituitary perfusion and intraventricular injection of catecholamines on prolactin release. *Endocrinology, 88:* 1012-1020.

Kamberi, I.A., Mical, R.S., and Porter, J.C. (1971b): Effects of melatonin and serotonin on the release of FSH and prolactin. *Endocrinology, 88:* 1288-1293.

Karobath, M., and Leitich, H. (1974): Antipsychotic drugs and dopamine-stimulated adenylate cyclase prepared from corpus striatum of rat brain. *Proceedings of the National Academy of Sciences, 71:* 2915-2918.

Kebabian, J.W., Petzhold, G.L., and Greengard, P. (1972): Dopamine sensitive adenylate cyclase in the caudate of rat brain and its similarity to the "dopamine receptor." *Proceedings of the National Academy of Sciences, 69:* 2145-2149.

Kehr, W., Carlsson, A., Lindqvist, M., Magnusson, T., and Atack, C.V. (1972): Evidence for a receptor-mediated feedback control of striatal tyrosine hydroxylase activity. *Journal of Pharmacy and Pharmacology, 24:* 744-747.

Keller, H.H., Bartholini, G., and Pletscher, A. (1973): Increase of 3-methoxy-4-hydroxyphenylethylene glycol in rat brain by neuroleptic drugs. *European Journal of Pharmacology, 23:* 183-186.

Kendell, R.E. (1975): What are our criteria for a diagnosis of schizophrenia? In: *On the Origin of Schizophrenic Psychoses.* Edited by H.M. van Praag. Erven Bohn, Amsterdam. Pp. 125-137.

Kety, S.S. (1967): Current biochemical approaches to schizophrenia. *New England Journal of Medicine, 276:* 325-331.

Kety, S.S. (1975): Mental illness in the biological and adoptive families of adopted individuals who have become schizophrenic. In: *On the Origin of Schizophrenic Psychoses.* Edited by H.M. van Praag. Erven Bohn Amsterdam. Pp. 19-26.

Kleinberg, D.L., Noel, G.L., and Frantz, A.G. (1971): Chlorpromazine stimulation and L-DOPA suppression of plasma prolactin in man. *Journal of Clinical Endocrinology and Metabolism, 33:* 873-876.

Korf, J., Praag, H.M. van and Sebens, J.B. (1971): Effect of intravenously administered probenecid in humans on the levels of 5-hydroxyindoleacetic acid, homovanillic acid and 3-methoxy-4-hydroxy-phenyl-glycol in cerebrospinal fluid. *Biochemical Pharmacology,* 20:659-668.

Korf, J., Praag, H.M. van, Schut, T., Nienhuis, R.J., and Lakke, J.P.W.F. (1974): Parkinson's disease and amine metabolites in cerebrospinal fluid: implications for L-DOPA therapy. *European Neurology, 12:* 340-350.

Kraepelin, E. (1896): *Psychiatrie. Ein Lehrbuch für Studierende und Arzte.* Barth, Leipzig. 5th ed.

Kraepelin, E. (1899): *Psychiatrie. Ein Lehrbuch für Studierende und Arzte.* Barth, Leipzig. 6th ed.

Kutzbach, C., and Stokstad, E.L.R. (1967): Feedback inhibition of methylenetetrahydrofate reductase in rat liver by S-adenosyl-methionine. *Biochimica et Biophysica Acta, 139:* 217-220.

Laduron, P. (1972): N-methylation of dopamine to epinine in brain tissue using N-methyltetrahydrofolic acid as the methyl donor. *Nature/New Biology, 238:* 212-213.

Laduron, P., Gommeron, W. and Leysen, J. (1974a): N-methylation of biogenic amines. *Journal of Biochemical Pharmacology, 23:* 1599-1608.

Laduron, P. (1974b): A new hypothesis on the origin of schizophrenia. *Journal of Psychiatric Research, 11:* 257-258.

Laduron, P. (1975): Scope and limitation in dopamine: β-hydroxylase measurement. *Biochemical Pharmacology, 24:* 557-562.

Lakke, J.P.W.F., Korf, J., Praag, H.M. van, and Schut, T. (1972): Predictive value of the probenecid test for the effect of L-DOPA therapy in Parkinson's disease. *Nature/New Biology, 236:* 208-209.

Langfeldt, G. (1939): *The Schizophreniform States.* Oxford University Press, London.

Lamprecht, F., Wyatt, R.J., Belmaker, R., Murphy, D.L., and Pollin, W. (1973): Plasma dopamine-β-hydroxylase activity in identical twins discordant for schizophrenia. In: *Frontiers in Catecholamine Research.* Edited by S.H. Snyder and E. Usdin. Pergamon Press, New York. Pp. 1123-1126.

Laverty, R., and Sharman, D.F. (1965): Modification by drug of the metabolism of 3,4-dihydroxyphenylethylamine, noradrenaline and 5-hydroxytryptamine in the brain. *British Journal of Pharmacology, 24:* 759-772.

Levi, R., and Waxman, S. (1975): Schizophrenia, epilepsy, cancer, methionine, and folate metabolism. *Lancet, II:* 11-13.

Leysen, J., and Laduron, P. (1974): N-methylation of indolealkylamines in the brain with a new methyl donor. In: *Advances in Biochemical Psychopharmacology,* vol. 11. Edited by E. Costa, G.L. Gessa and M. Sandler. Raven Press, New York. Pp. 65-93.

Libertun, C., and McCann, S.M. (1973): Blockade of the release of gonadotropins and prolactin by subcutaneous or intraventricular injection of atropine in male and female rats. *Endocrinology, 92:* 1714-1724.

Lu, K-H., Amenomori, Y., Chen, C-L., and Meites, J. (1970): Effect of central acting drugs on serum and pituitary prolactin levels in rats. *Endocrinology, 87:* 667-672.

Lu, K-H., and Meites, J. (1971): Inhibition of L-DOPA and monoamine oxidase inhibitors of pituitary prolactin release, stimulation by methyldopa and d-amphetamine. *Proceedings of the Society for Experimental Biology and Medicine, 137:* 480-483.

Lu, K-H., and Meites, J. (1973): Effects of serotonin precursors and melatonin on serum prolactin release in rats. *Endocrinology, 93:* 152-155.

Mandell, A.J., and Morgan, M. (1971): Indole(ethyl)amine N-methyltransferase in human brain. *Nature/New Biology, 230:* 85-87.

Matthysse, S. (1973): Antipsychotic drug actions: a clue to the neuropathology of schizophrenia? *Federation Proceedings, 32:* 200-205.

Matthysse, S. (1974): Implications of catecholamine systems of the brain in schizophrenia. In. *Brain Disfunction in Metabolic Disorders*. Edited by F. Plum. Raven Press, New York. Pp. 305-315.

Mayer-Gross, W., Slater, F., and Roth, M. (1960): *Clinical Psychiatry*. Williams and Wilkins, Baltimore.

Meites, J., Lu, K-H., Wuttke, W., Welsch, C.W., Nagasawa, H., and Quadrie, F.K. (1972): Recent studies on functions and control of prolactin secretion in rats. *Recent Progress in Hormone Research, 28:* 471-526.

Meller, E., Rosengarten, H., Friedhoff, A.J., Stebbins, R.D., and Silber, R. (1975): 5-Methyltetrahydrofolic acid is not a methyl donor for biogenic amines: enzymatic formation of formaldehyde. *Science, 187:* 171-173.

Mellor, C.S. (1970): First rank symptoms of schizophrenia: I. The frequency in schizophrenics on admission to hospital. II. Differences between individual first rank symptoms. *British Journal of Psychiatry, 117:* 15-23.

Meltzer, H., and Stahl, S.M. (1974): Platelet monoamine oxidase activity and substrate preferences in schizophrenic patients. *Rexearch Communications in Chemical Pathology and Pharmacology, 7:* 419-431.

Meltzer, H., Sachar, E.J., and Frantz, A.G. (1974): Serum prolactin levels in unmedicated schizophrenic patients. *Archives of General Psychiatry, 31:* 564-569.

Meltzer, H., Sachar, E.J., and Frantz, A.G. (1975): Dopamine antagonism by thioridazine in schizophrenia. *Biological Psychiatry, 10:* 53-57.

Meltzer, H., Sachar, E.J., and Frantz, A. (in press): *Prolactin Secretion in Schizophrenia*.

Mendels, J. (1975): *Psychobiology of Depression*. Halsted Press, New York.

Miller, R.J., Horn, A.S., and Iversen, L.L. (1974): The action of neuroleptic drugs in dopamine stimulated adenosine-3',5'-monophosphate production in rat neostriatum and limbic forebrain. *Molecular Pharmacology, 10:* 759-766.

Morel, B. (1860): *Traité des maladies mentales*. Masson, Paris.

Murphy, D.L., and Wyatt, R.J. (1972): Reduced monoamine oxidase activity in blood platelets from schizophrenic patients. *Nature, 238:* 225-226.

Nagatsu, T., Hidaka, H., Kuzuya, H., and Takeya, K. (1970): Inhibition of dopamine-β-hydroxylase by fusaric acid (5-butylpicolinic acid) in vitro and in vivo. *Biochemical Pharmacology, 19:* 35-44.

Narasimhachari, N., Heller, B., Spaide, J., Hascovec, L., Fujimori, M., Tabushi, K., and Himwich, H.E. (1970): Comparative behavioral and biochemical effects of tranylcypromine and cysteine on normal controls and schizophrenic patients. *Life Sciences, 9:* 1021-1032.

Narasimhachari, N., Heller, B., Spaide, J., Hascovec, L., Meltzer, H., Strahilevitz, M., and Himwich, H.E. (1971): N,N-dimethylated indoleamines in blood. *Biological Psychiatry, 3:* 21-23.

Nies, A., Robinson, B.S., Harris, L.S., and Lamborn, K.L. (1974): Comparison of monoamine oxidase, substrate activities in twins, schizophrenics, depressives and controls. In: *Advances in Biochemical Psychopharmacology*, vol. 12. Edited by E. Usdin. Raven Press, New York. Pp. 59-71.

Nybäck, H., Borzecki, Z., and Sedvall, G. (1968): Accumulation and disappearance of catecholamines formed from tyrosine-^{14}C in mouse brain; effect of some psychotropic drugs. *European Journal of Pharmacology, 4:* 395-403.

Nybäck, H., and Sedvall, G. (1970): Further studies on the accumulation and disappearance of catecholamines formed from tyrosine-^{14}C in mouse brain. Effect of some phenothiazine analogues. *European Journal of Pharmacology, 10:* 193-205.

Nybäck, H., Schubert, J., and Sedvall, G. (1970): Effect of apomorpine and pimozide on synthesis and turnover of labelled catecholamines in mouse brain. *Journal of Pharmacy and Pharmacology, 22:* 622-624.

O'Keefe, R., Sharman, D.F., and Vogt, M. (1970): Effect of drugs used in psychoses on cerebral dopamine metabolism. *British Journal of Pharmacology, 38:* 287-304.

Olson, L. (1974). Post-mortem fluorescence histochemistry of monoamine neuron systems in the human brain: a new approach in the search for a neuropathology of schizophrenia. *Journal of Psychiatric Research, 11:* 199-203.

Olsson, R., and Roos, B-E. (1968): Concentrations of 5-hydroxyindoleacetic acid and homovanillic acid in the cerebrospinal fluid after treatment with probenecid in patients with Parkinson's disease. *Nature, 219:* 502-503.

Osmond, H., and Smythies, J. (1952): Schizophrenia: a new approach. *Journal of Mental Sciences, 98:* 309-315.

Park, D.C., Baldessarini, R.J., and Kety, S.S. (1965): Methionine effects on chronic schizophrenics. *Archives of General Psychiatry, 12:* 346-351.

Perry, T.L., Hansen, S., and Kloster, M. (1973): Huntington's chorea: deficiency of γ-amino-butyric acid in brain. *New England Journal of Medicine, 288:* 337-342.

Persson, T., and Roos, B-E. (1969): Acid metabolites from monoamines in CSF of chronic schizophrenics. *British Journal of Psychiatry, 115:* 95-98.

Pieri, L., Pieri, M., and Haefely, W. (1974): LSD as an agonist of dopamine receptors in the striatum. *Nature, 252:* 586-589.

Pletscher, A., Gey, K.F., and Burkard, W.P. (1967): Effect of neuroleptics on the cerebral metabolism of catecholamines. In: *Proceedings of the European Society for the Study of Drug Toxicity,* vol. 9. Toxicity and Side Effects of Psychotropic Drugs. Excerpta Medica International Congress, Paris. Pp. 98-106.

Pollard, H.B., Barker, J.L., Bohr, W.A., and Dowdall, M.J. (1975): Chlorpromazine: specific inhibition of l-noradrenaline and 5-hydroxytryptamine uptake in synaptosomes from squid brain. *Brain Research, 85:* 23-31.

Pollin, W., Cardon, P.V., and Kety, S.S. (1961): Effects of amino acid feeding in schizophrenic patients treated with iproniazid. *Science, 133:* 104.

Porter, J.C., Mical, R.S., and Cramer, O.M. (1971): Effect of serotonin and other indoles on the release of LH, FSH, and prolactin. *Horm. Antagon. Gynecol. Invest., 2:* 13-22.

Post, R.M., Fink, E., Carpenter, W.T., and Goodwin, F.K. (1975): Cerebrospinal fluid amine metabolites in acute schizophrenia. *Archives of General Psychiatry, 32:* 1063-1069.

Post, R M., and Goodwin, F.K. (1975): Time-dependent effects of phenothiazines on dopamine turnover in psychiatric patients. *Science, 190:* 488-489.

Praag, H.M. van (1962): *The significance of Monoamine Oxidase Inhibition and the Therapeutic Principle in the Treatment of Depression.* Thesis, Utrecht.

Praag, H.M. van, and Leijnse, B. (1963): Die Bedeutung der Psychopharmakologie für die klinische Psychiatrie. Systematik als notwendiger Ausgangspunkt. *Nervenarzt, 34:* 530-537.

Praag, H.M. van, and Leijnse, B. (1965): Neubewertung des Syndroms. Skizze einer funktionellen Pathologie. *Psychiatria, Neurologia, Neurochirurgia, 68:* 50-66.

Praag, H.M. van (1967): The possible significance of cerebral dopamine for neurology and psychiatry. *Psychiatria, Neurologia, Neurochirurgia, 70:* 361-379.

Praag, H.M. van, and Korf, J. (1971): Endogenous depressions with and without disturbances in the 5-hydroxytryptamine metabolism: a biochemical classification? *Psychopharmacologia, 19:* 148-152.

Praag, H.M. van (1974a): Towards a biochemical typology of depressions? *Pharmacopsychiatry, 7:* 281-292.

Praag, H.M. van (1974b): New development in human psychopharmacology. *Comprehensive Psychiatry, 15:* 389-401.

Praag, H.M. van (1975): Neuroleptics as a guideline to biological research in psychotic disorders. *Comprehensive Psychiatry, 16:* 7-22.

Praag, H.M. van, Korf, J., Lakke, J.P.W.F., and Schut, T. (1975a): Dopamine metabolism in depressions, psychoses, and parkinson's disease: the problem of the specificity of biological variables in behavior disorders. *Psychological Medicine, 5:* 138-146.

Praag, H.M. van, Dols, L.C.W., and Schut, T. (1975b): Biochemical versus psychopathological action profile of neuroleptics: A comparative study of chlorpromazine and oxypertine in acute psychotic disorders. *Comprehensive Psychiatry, 16:* 255-263.

Praag, H.M. van, and Korf, J. (1975): Neuroleptics, catecholamines, and psychoses: a study of their interrelations. *American Journal of Psychiatry, 132:* 593-597.

Praag, H.M. van, and Korf, J. (1976a): The importance of dopamine in the pathogenesis of psychosis and the action of antipsychotic (neuroleptic) drugs. Paper read at the Sixth International Congress of Pharmacology, Helsinki, July 1975.

Praag, H.M. van, and Korf, J. (1976b): Importance of the dopamine metabolism for the clinical effects and side effects of neuroleptics. *American Journal of Psychiatry.* (in press)

Praag, H.M. van, Korf, J., and Lequin, R.M. (1976): An unexpected effect of 1-5-hydroxy-tryptophan-ethyl-ester combined with a peripheral decarboxylase inhibitor on human serum prolactin. *Psychopharmacology Communications.* In press.

Praag, H.M. van (1977a): The significance of dopamine for the mode of action of neuroleptics and the pathogenesis of schizophrenia. *British Journal of Psychiatry.* In press.

Praag, H.M. van (1977b): Significance of biochemical parameters in the diagnosis, treatment and prevention of depressive disorders. *Biological Psychiatry.* In press.

Praag, H.M. van, Korf, J., and Dols, L.C.W. (1977): Clozapine versus perphenazine or, the value of the biochemical mode of action of neuroleptics in predicting their therapeutic activity. *British Journal of Psychiatry.* In press.

Praag, H.M. van (1977): Satisfaction and disappointment. Reflections at an anniversary. In: *Neurotransmission and Disturbed Behaviour.* Edited by H.M. van Praag. Bohn Scheltema & Holkema, Utrecht. In press.

Rackensperger, W., Gaupp, R., Mattke, D.J., Schwartz, D., and Stutte, K.H. (1974): Behandlung von akuten schizophrenen Psychosen mit Beta-Receptoren-Blockern. *Archiv für Psychiatrie und Nervenkrankheiten, 219:* 29-36.

Randrup, A., and Munkvad, I. (1966): Role of catecholamines in the amphetamine exicatory response. *Nature, 211:* 540.

Randrup, A., and Scheel-Krüger, J. (1966): Diethyldithiocarbonate and amphetamine stereotype behavior. *Journal of Pharmacy and Pharmacology, 18:* 752.

Randrup, A., and Munkvad, I. (1970): Biochemical, anatomical and psychological investigations of stereotyped behavior induced by amphetamines. In: Amphetamines and related compounds. Edited by E. Costa and S. Garattini. Raven Press, New York. Pp. 695-713.

Randrup, A., and Munkvad, I. (1972): Evidence indicating an association between schizophrenia and dopaminergic hyperactivity in the brain. *Orthomolecular Psychiatry, 1:* 2-27.

Rech, R.H., Borys, H.K., and Moore, K.E. (1966): Alterations in behavior and brain catecholamine levels in rats treated with α-methyltyrosine. *Journal of Pharmacology and Experimental Therapeutics, 153:* 412-419.

Reich, W. (1975): The spectrum concept of schizophrenia. *Archives of General Psychiatry, 32:* 489-498.

Reid, A.A. (1973): Schizophrenia—Disease or Syndrome? *Archives of General Psychiatry, 28:* 863-869.

Richardson, J.S., and Jacobowitz, D.M. (1973): Depletion of brain norepinephrine by intraventricular injection of 6-hydroxydopa: a biochemical, histochemical and behavioral study in rats. *Brain Research, 58:* 117-135.

Rimón, R., Roos, B-E., Räkköläinen, V., and Alanen, Y. (1971): The content of 5-hydroxy-indoleacetic acid and homovanillic acid in the cerebrospinal fluid of patients with acute schizophrenia. *Journal of Psychosomatic Research, 15:* 375-378.

Kobins, E., and Guze, S.B. (1970): Establishment of diagnostic validity in psychiatric illness: its applications to schizophrenia. *American Journal of Psychiatry, 126:* 983-987.

Rollema, H., Westerink, B.H.C., and Grol, C.J. (1976): Correlation between neuroleptic induced suppression of stereotyped behavior and HVA levels in the rat brain. *J Pharm. Pharmacol.* 321-323.

Roos, B-E. (1965): Effects of certain tranquillizers on the level of homovanillic acid in the corpus striatum. *Journal of Pharmacy and Pharmacology, 17:* 820-821.

Rosenblatt, S., Leighton, W.P., and Chanley, J.D. (1973): Dopamine-β-hydroxylase: evidence for increased activity in sympathetic neurons during psychotic states. *Science, 182:* 923-924.

Rosenthal, D. (1970): *Genetic Theory and Abnormal Behavior.* McGraw-Hill Book Company, New York.

Rossum, J.M. van (1967): The significance of dopamine receptor blockade for the action of neuroleptic drugs. In: *Neuropsychopharmacology.* Edited by H. Brill. Excerpta Medica Foundation, Den Haag. Pp. 321-329.

Rümke, H.C. (1967): Uber die Schizophrenie. In: *Eine blühende Psychiatrie in Gefahr.* Edited by H.C. Rümke. Springer, Berlin. Pp. 201-242.

Rylander, G. (1972): Psychoses and the punding and choreiform syndromes in addiction of central stimulant drugs. *Psychiatria, Neurologia and Neurochirurgia, 75:* 203-212.

Saavedra, J.M., and Axelrod, J. (1972): Psychotomimetic N-methylated tryptamines: formation in brain in vivo and in vitro. *Science, 175:* 1365-1366.

Saavedra, J.M., Coyle, J.T., and Axelrod, J. (1973): The distribution and properties of the nonspecific N-methyltransferase in brain. *Journal of Neurochemistry, 20:* 743-752.

Sack, R.L., and Goodwin, F.K. (1974): Inhibition of dopamine-β-hydroxylase in manic patients. *Archives of General Psychiatry, 31:* 649-654.

Sathanathan, G., Angrist, B.M., and Gershon, S. (1973): Response threshold to L-DOPA in psychiatric patients. *Biological Psychiatry, 7:* 139-146.

Scales, B., and Cosgrove, M.B. (1970): The metabolism and distribution of the selective adrenergic beta blocking agent, practolol. *Journal of Pharmacology and Experimental Therapeutics, 175:* 338-347.

Schally, A.V., Arimura, A., and Kastin, A.J. (1973): Hypothalamic regulatory hormones. *Science, 179:* 341-350.

Schneider, K. (1959): *Klinische Psychopathologie.* Georg Thieme Verlag, Stuttgart. Fünfte Auflage.

Schwartz, M.A., Wyatt, R.J., Yang, H.Y., and Neff, N.H. (1974): Multiple forms of brain monoamine oxidase in schizophrenic and normal individuals. *Archives of General Psychiatry, 31:* 557-560.

Sedgwick, P. (1975): The social analysis of schizophrenia. Theories of Laing, Cooper, etc. In: *On the origin of Schizophrenic Psychoses.* Edited by H.M. van Praag. Erven Bohn, Amsterdam. Pp. 183-208.

Sedvall, G., Bjerkenstedt, L., Fyrö, B., Härnryd, C., Wiesel, F.A., and Wode-Helgodt, B. (1976): Selective effects of psychoactive drugs on levels of monoamine metabolites in lumbar cerebrospinal fluid of psychiatric patients. Paper read at the Sixth International Congress of Pharmacology, Helsinki.

Seeman, P., Chau-Wong, M., Tedesco, J., and Wong, K. (1975): Brain receptors for antipsychotic drugs and dopamine: Direct binding assays. *Proceedings of the National Academy of Sciences, 72:* 4376-4380.

Seeman, P., and Lee, T. (1975): Antipsychotic drugs: direct correlation between clinical potency and presynaptic action on dopamine neurons. *Science, 188:* 1217-1219.

Segal, D.S., and Mandell, A.J. (1970): Behavioural activation of rats during intraventricular infusion of norepinephrine. *Proceedings of the National Academy of Sciences, 66:* 289-293.

Serban, G., and Gidynski, C.B. (1975): Differentiating criteria for acute-chronic distinction in schizophrenia. *Archives of General Psychiatry, 32:* 705-712.

Shaskan, G.E., and Becker, R.E. (1975): Platelet monoamine oxidase in schizophrenics. *Nature, 253:* 659-660.

Shopsin, B., Freedman, L.S., and Goldstein, M. (1972): Serum dopamine-β-hydroxylase (DβH) activity and affective states. *Psychopharmacologia, 27:* 11-16.

Singh, M.M., and Kay, S.R. (1975): Therapeutic reversal with benztropine in schizophrenics. Practical and theoretical significance. *Journal of Nervous and Mental Disease, 160:* 258-266.

Smith, R.C., Boggan, W.O., and Freedman, D.X. (1975): Effects of single and multiple dose LSD on endogenous levels of brain tyrosine and catecholamines. *Psychopharmacologia, 42:* 271-276.

Smythies, J.R. (1975): The biochemical basis of schizophrenia. In: *New Perspectives in Schizophrenia.* Edited by A. Forrest and J. Affleck. Churchill Livingstone, Edinburgh. Pp. 51-68.

Snyder, S.H., Taylor, K.M., Coyle, J.R., and Meyerhoff, J.L. (1970): The role of brain dopamine in behavioral regulation and the actions of psychotropic drugs. *American Journal of Psychiatry, 127:* 199-207.

Snyder, S.H., Aghajanian, G.K., and Matthysse, S. (1972): Pharmacological observations: drug induced psychoses: effects of psychotropic drugs on aminergic systems. *Neurosci. Res. Prog. Bull., 10:* 430-445.

Snyder, S.H. (1973): Amphetamine psychosis: a "model" schizophrenia mediated by catecholamines. *American Journal of Psychiatry, 130:* 61-66.

Snyder, S.H., Banerjee, S.P., Yamamura, H.I., and Greenberg, D. (1974): Drugs, neurotransmitters and schizophrenia. *Science, 184:* 1243-1253.

Sourkes, T.L., and Poirier, L.J. (1966): Neurochemical bases of tremor and other disorders of movement. *Canadian Medical Association, 94:* 53-60.

Southgate, P.J., Mayer, S.R., Boxall, E., and Wilson, A.B. (1971): Some 5-hydroxytryptamine-like actions of fenfluramine: a comparison with (+)-amphetamine and diethylpropion. *Journal of Pharmacy and Pharmacology, 23:* 600-605.

Stein, L., and Wise, C.D. (1971): Possible etiology of schizophrenia: progressive damage to the noradrenergic reward system by 6-hydroxydopamine. *Science, 171:* 1032-1036.

Steiner, M., Latz, A., Blum, I., Atsmon, A., and Wijsenbeek, H. (1973): Propranolol versus chlorpromazine in the treatment of psychoses associated with childbearing. *Psychiatria, Neurologia and Neurochirurgia, 76:* 421-426.

Strömgren, E. (1975): Genetic factors in the origin of schizophrenia. In: *On the Origin of Schizophrenic Psychoses.* Edited by H.M. van Praag. Erven Bohn, Amsterdam. Pp. 7-18.

Svensson, T.H., and Waldeck, B. (1970): On the role of brain catecholamines in motor activity: experiments with inhibitors of synthesis and monoamine oxidase. *Psychopharmacologia, 18:* 357-365.

Tanimukai, H., Ginther, R., Spaide, J., Bueno, J.R., and Himwich, H.E. (1967): Occurrence of bufotenine in urine of schizophrenic patients. *Life Sciences, 6:* 1697-1706.

Tanimukai, H., Ginther, R., Spaide, J., Bueno, J.R., and Himwich, H.E. (1970): Detection of psychotomimetic N,N-dimethylated indoleamines in the urine of four schizophrenic patients. *British Journal of Psychiatry, 117:* 421-430.

Taylor, M. (1972): Schneiderian first-rank symptoms and clinical prognostic features in schizophrenia. *Archives of General Psychiatry, 26:* 64-67.

Thoenen, H., Hürlimann, A., and Haefely, W. (1965): On the mode of action of chlorpromazine on peripheral adrenergic mechanisms. *International Journal of Neuropharmacology, 4:* 79-89.

Thoenen, H., and Tranzer, J.P. (1973): The pharmacology of 6-hydroxydopamine. *Annual Review Pharmacology, 13:* 169-180.

Westerink, B.H.C., and Korf, J. (1975a): Influence of drugs on striatal and limbic homovanillic acid concentration in the rat brain. *European Journal of Pharmacology, 33:* 31-40.

Westerink, B.H.C., and Korf, J. (1975b): Determination of nanogram amounts of homovanillic acid in the central nervous system with a rapid semiautomated fluorometric method. *Biochemical Medicine, 12:* 106-114.

Wiesel, F.A., and Sedvall, G. (1975): Effect of antipsychotic drugs on homovanillic acid levels in striatum and olfactory tubercle of the rat. *European Journal of Pharmacology, 30:* 364-367.

Wise, C.D., and Stein, L. (1973): Dopamine-beta-hydroxylase. Deficits in the brains of schizophrenic patients. *Science, 181:* 344-347.

Wise, D., Stein, L. (1975): Discussion remarks. *Science, 187:* 370.

Woodruff, G.N. (1971): Dopamine receptors: a review. *Comparative and General Pharmacology, 2:* 439-441.

Wyatt, R.J., Murphy, D.L., Belmaker, R., Cohen, S., Donelley, C.H., and Pollin, W. (1973): Reduced monoamine oxidase activity in platelets: a possible genetic marker for vulnerability to schizophrenia. *Science, 179:* 916-918.

Wyatt, R.J., Schwartz, M.A., Erdelyi, J.D., and Barchas, J.D. (1975): Dopamine-β-hydroxylase activity in brains of chronic schizophrenic patients. *Science, 187:* 368-370.

Yaryuara-Tobias, J.A., Diamond, B., and Merlis, S. (1970): The action of L-DOPA on schizophrenic patients (a preliminary report). *Current Therapeutic Research, 12:* 528-531.

Yorkston, N.J., Zaki, S.A., Malik, M.K.U., Morrison, R.C., and Havard, C.W.H. (1974): Propranolol in the control of schizophrenic symptoms. *British Medical Journal, 4:* 633-635.

Youdim, M.B.H., Woods, H.F., Mitchell, B., Grahame-Smith, D.G., and Callender, S. (1975): Human platelet monoamine oxidase activity in iron-deficiency anaemia. *Clinical Science and Molecular Medicine, 48:* 289-295.

Young, D., and Scoville, W.B. (1938): Paranoid psychosis in narcolepsy and the possible dangers of benzedrine treatment. *Medical Clinics of North America, 22:* 637-646.

Zeller, E.A., Boshes, B., Davis, J.M., and Thorner, M. (1975): Molecular aberration in platelet monoamine oxidase in schizophrenia. *Lancet, I:* 1385.

Subject Index

PENGUIN BOOKS

ANDREW ROBERTS

Churchill
Walking with Destiny

PENGUIN BOOKS

UK | USA | Canada | Ireland | Australia
India | New Zealand | South Africa

Penguin Books is part of the Penguin Random House group of companies
whose addresses can be found at global.penguinrandomhouse.com.

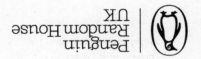

First published by Allen Lane 2018
Published in Penguin Books 2019
006

Copyright © Andrew Roberts, 2018
The moral right of the author has been asserted

Set in 8.78/11.6 pt Sabon LT Std
Typeset by Jouve (UK), Milton Keynes
Printed and bound in Great Britain by Clays Ltd, Elcograf S.p.A.

A CIP catalogue record for this book is available from the British Library

ISBN: 978-0-141-98125-3

www.greenpenguin.co.uk